Frederick Brendel

Flora Peoriana

The vegetation in the climate of middle Illinois

Frederick Brendel

Flora Peoriana
The vegetation in the climate of middle Illinois

ISBN/EAN: 9783337268619

Printed in Europe, USA, Canada, Australia, Japan

Cover: Foto ©berggeist007 / pixelio.de

More available books at **www.hansebooks.com**

Flora Peoriana

THE VEGETATION IN THE CLIMATE

OF MIDDLE ILLINOIS

BY FREDERICK BRENDEL

PEORIA, ILL.:

J. W. FRANKS & SONS, PRINTERS AND BINDERS

1887

PREFACE.

The contents of this essay is the result of thirty-five years observations on the vegetation of a small area of about three hundred English square miles. It is intended to show how local floras should be treated to be useful to phytogeography; how notice should be taken of soil and climate, to understand the vegetation of a certain floral district. Descriptions of genera and species are unnecessary, in such treatise, as those can be found in Gray's Manual, which, being in every botanists hand, offers sufficient means to identify all the species found in our country.

May this little book take the initiative to similar treatises, that would in a great measure, facilitate the work of a future writer on phytogeography of North America.

ERRATA.

Page 17, line 11 (from bottom), for 15 read _1.5.

Page 20, line 15 (from top), for tribola read triloba.

Page 20, line 16 (from top), for June 17 read June 7.

Page 22, line 9 (from bottom), for 27.560 read 29.560.

Page 25, table, April, column 6, for 11 read 12.

Page 25, table, June, column 9, for 315 read 316.

Page 25, table, August, column 5, for 4 read 3.

Page 25, table, October, column 3, for 35 read 36.

Page 27, table 8, line 17, for fall read year.

Page 35, line 21. for Sagifraga read Saxifraga.

Page 36, line 17, for Saggittaria read Sagittaria.

Page 38, line 12 (from bottom). for Ipomoce read Ipomoea.

Page 40, line 24, for Condolle read Candolle.

Page 43, line 18, after VIII., insert 5.

Page 43, line 22, for All. read Atl.

Page 51, line 30, for Eschinacea read Echinacea.

Page 54, line 5, for S read 5.

Page 70, last line, for Orhiglossum read Ophioglossum.

GENERAL REMARKS ON DISTRIBUTION OF PLANTS.

When we study the history of a country, we ought to be acquainted with its geography, its physiognomy of the landscape, its climate and the physical qualities of its people. All these things will influence the moral character of the people and only in that way, combining cause and effect, we will gain a clear view of its history.

Likewise when we study the flora of a country, it is not sufficient to know the names and characteristic qualities of all the species that grow in a certain district; we ought to know the circumstances under which they grow, the topography, the climate, the nature of the soil and the geographical distribution of each species beyond the limits of the country in question.

This branch of science, known by the name of phytogeography, is a comparatively new one; it was founded in the first decennium of our century by Alexander Humboldt, when he published, in 1805, his "Essai sur la Géographie des Plantes," and in 1817, his "Prolegomena de Distributione Geographica Plantarum."

Since that time many botanists, by treatises on single countries or on single groups of plants, have furnished material to more general works on the subject, f. i., Wahlenberg on the flora of Lapponia, 1812: on the vegetation and climate of Northern Switzerland, 1813; in his Flora Carpathorum, 1814. Robt. Brown, in his "General Remarks, Geographical and Systematical on the Botany of Terra Australis, 1814, etc.

The first attempt to arrange the vegetation of our globe into separate geographical divisions was made by the Danish botanist Schouw, in 1822, when he published "Grundtrack til en almindelig Plantegeographie," followed in 1824 by an Atlas of twenty-two maps. He used the names of the characteristic orders for each of his divisions: For North America, three only; the most northern, from 50 N. L. northward, he called the Kingdom of Saxifrageæ and Mosses; the Northern United States and Canada formed that of the Asters and Solidagines; the Southern States that of the Magnolias. The country between the Rocky Mountains and the Pacific Coast, the flora of which was at that time nearly unknown, he passed by. Meyen, the botanist of the Prussian expedition around the world in 1830-32, published in 1836 a general work on the Geography of Plants. The most important works on the subject are the "Géographie Botanique," by Alph. De Candolle (1855), and "Grisebach's Vegetation of the Earth," in 1872.

Grisebach's botanical provinces are much more natural than Pickering's quite artificial divisions in U. S. Expl. Expeditions, XV; but in regard to the causes that effected the distinct floras, he is rather one-sided; excluding all geological causes, admitting of recent agencies of migration only, and refuting the theory of transmutation of species, he adheres to the obsolete belief in general revolutions of the globe and new creations. As some other prominent scientists, he seems to be influenced by hereditary religious prejudices. However his arrangement of botanical provinces may partly be retained, as far as new discoveries in less thoroughly examined countries will not demand some changes, as proposed by A. Engler, in his Essay on a History of Evolution of the Vegetable World (1879–82), and by O. Drude (Florareiche der Erde 1884), who both differ from Grisebach in their argumentation acknowledging transmutation of species and geological agencies.

The limitation, not only of species but of whole genera, and even orders within continents or certain parts of the same, created the idea of a plurality of primitive centres of creation and of a phytogeographical significance of endemismus and of monotypes. This view, and the denial of a genetical connection of the species through all geological periods agrees perfectly well with a dualistic conception of cosmogony, the hypothesis of an arbitrary power, that created highly organized beings by an immediate will and placed each kind at a certain locality of our globe.

A more probable hypothesis is the modern monistic biological theory: By an unknown compulsive force are uninterruptedly produced new forms out of older ones, and more perfect ones, beginning with the most simple organizations, so that the recent forms, by convenience called species, are genetically connected with the extinct.

Species become extinct in certain localities and are preserved in others; only that way we can understand the co-existence of one and the same species in widely disjoint countries, f. i., of our Phryma leptostachya in North America and the Himalayas; for nobody would believe that this plant has a double origin, and the only explanation of this fact is that the plant had formerly a wider distribution, and became extinct in countries between the actual habitats. Several species of Liriodendron are found in the Miocene formation of Greenland, and, besides, in Germany and Italy, when now only one species exist in North America; so the Taxodium distichum, another North American tree, was found in the bituminous slate-clay of Spitzbergen. Therefore, quite properly Bentham proposed, instead of centre of creation, the term centre of preservation.

The above mentioned species show that endemismus has nothing to do with origin, but only preservation, and that only in that sense it is of any value in phytogeography. Monotypes are very often the arbitrary make of systematics, and depend of their proneness to narrower or wider limitation of species and genera. Hepatica triloba is a monotype as soon as we separate it from Anemone, and as soon as we unite with it the other little founded species. But when we acknowledge the latter as " good " species, then it ceases to be a monotype. Pentachæta was, when Nuttall proposed the genùs, a monotype until A. Gray described a second species, and with that as a variety, another monotype: Aphantochæta. So both of them cease to be monotypes.

Of eight other genera of the order Compositæ, which Grisebach mentions as monotypes (in Vegetation of the Earth), only two : Whitneya and Crocidium retain yet their monotypism: Actinolepris merged in Eriophyllum, Oxyura in Layia, Coinogyne in Iaumea, Tuckermannia in Leptosyne. Corethrogyne is now represented by three Hulsea by six species, and besides two varieties. In the meantime not less than ten California monotypes are proposed, the greater part, probably, waiting for the company of new foundlings.

These few examples, out of many, will show the value of monotypes the more when we add the above mentioned Phryma, which is not only a monotype genus, but after Schauer, even a monotype order, and does exist in two so remote countries.

In undisputed monotypes what else can we recognize but the isolated remnants of an extinct plurality of forms, isolated by a row of geological and climatological changes? Analogous examples are offered in Zoology, when we compare the small number of recent ganoids with their abundance in early geological periods.

Comparatively few species of plants are distributed all over the surface of our globe, of which a large proportion again proved to have spread by migration, even in historical time. The great majority of species is restricted to certain areas within certain limits that are defined by climatical influences as heat, light, humidity, or by physical or chemical qualities of the soil, or by geographical obstacles as are oceans, deserts or high mountain ranges, which a species may be unable to pass over.

The assemblage of species which, under the above mentioned influences, grow in a part of our globe, giving to the same its characteristic physiognomy, we call the flora of that country and the area to which it, as a whole, is restricted, we call a botanical province with subordinate regions and districts. It is evident that political boundaries must be excluded; the flora of a state may be quite different in its parts, f. i., in Virginia the flora is different in the Atlantic slope, in the Alleghanies and in the Western slope, and belongs, accordingly, to three different botanical districts, as will be shown.

To arrange the vegetable world geographically into natural provinces and districts, the best guide will be the character of the landscape and the statistic proportion of the species, genera and orders. Climate and nature of soil will serve to elucidate the facts.

Species, genera and orders, may be limited by lines, and so may be determined the area of each, but it will not do for the complex of them. For these lines will, in many ways, cross each other, and so the character of the landscape will not abruptly change; species after species will disappear and new ones will present themselves.

Traveling from the Atlantic westward, we find at first deciduous-leaved trees and coniferæ; on the Alleghenies, the same, with an additional number of Ericaceæ; then west of the Alleghenies, in the Ohio Valley, deciduous-leaved woods; on the Upper Mississippi, the same, interspersed with prairies, which prevail west of the Mississippi, only the banks of the rivers being wooded, and farther westward the trees disappear entirely. The similar changes take place going from the north to the south.

An early publication on the geographical distribution of North American plants, is Barton's specimen of a geographical view of the trees and shrubs of North America (1809), of the newer ones the most important are: Richardson's Chapter on Woody Plants and Carices, in the second volume of his Arctic Searching Expedition (1850;) then Cooper's Paper in Smithsonian Reports (1858), on the Distribution of the Forests and Trees of North America; A. Gray's Statistics of the Flora of United States (1856); Gray and Hooker's: The Vegetation of the Rocky Mountains in Geological and Geographical Survey of the Territories (1881), and finally Sargent's Forest Trees of North America, in Vol. IX of the United States Census (1884.)

For North America, I propose the following floral divisions:

1. The Arctic Alpine.
2. The Woodland.
3. The Californian.
4. The Prairie.
5. As a part, or at least a transition to the West Indian Flora, that of South Florida.

1. The Arctic Flora, which is nearly the same as in the eastern continent, covers all the country northeast of a line drawn from the mouth of the Mackenzie to the Hudson's Bay, under 60° north latitude, and north of a line from the mouth of the Mackenzie to the northwest coast under 66° north latitude; then that part of Labrador that lies north of 59°, and the highest points of the Rocky Mountains, Sierra Nevada and White Mountains. Characteristic is the absence of any tree-growth. The whole Arctic Flora does contain not quite four hundred and fifty species of vascular plants, and many mosses and lichens.

2. The North American Woodland, comprising the greater part of the continent from the Atlantic to the northeast, from Labrador and North Florida to East Texas, Missouri and the Lower Saskatchawan, could be sub-divided in the following provinces: A sub-arctic province from Alaska to the Hudson's Bay, an uninterrupted forest of coniferæ, mostly Pinus alba and a few poplars, birch, alder and willows. A North Pacific Province: The coast from the peninsula Alaska, to the Oregon and the wooded part of the Rocky Mountains, have

more or less the same flora, with peculiar coniferæ and trees with deciduous leaves; southward, increasing in number of species and passing over into the Californian province. A Province of Mixed Forests: Coniferæ and leaved-woods, occupying the tract of land from Lake Winnipeg and the great lakes to Nova Scotia and New Jersey; and, southward, in the Alleghanies to Georgia. This should be divided in three districts: *(a)* Around the great lakes (Canadian) coniferæ prevailing; *(b)* coast from Nova Scotia to Delaware, North Atlantic, deciduous trees prevailing; *(c)* Alleghanies, coniferæ, with many Ericaceæ and some Magnoliaceæ. A Province of Deciduous-leaved Trees with two districts: *(a)* The Ohio Valley, with nearly exclusively deciduous-leaved trees; *(b)* the Upper Mississippi, with deciduous-leaved trees and prairies, the former, westward, decreasing in number of species. A Province of Evergreen-leaved Trees, between the Atlantic coast from Virginia to North Florida, and along the Gulf of Mexico to the Brazos, in Texas. This could be divided in an eastern and western district, but the number of species, common to both, is so great, and the character of the landscape so similar, that it is not advisable. The area is coincident with that of the tertiary formation.

3. The California Flora, from San Diego to the Columbia River, and west of the Sierra Nevada. These mountains, as well as the coast range, are covered with many peculiar species of coniferæ. The valley of the Sacramento and San Joaquin is sparsely wooded mostly with oaks, among which are several perennial-leaved species. In the coast range, north of San Francisco, the thick forest does consist mostly of the same species in one locality, and of another in the next *a. s. o.* Southward of San Francisco the forests grow thinner and thinner, with only a few coniferæ, and at Santa Barbara the mountains are nearly bare or covered with low shrubs. There is the transition to the North Mexican Flora.

4. The region of the prairies is characterized by a dry atmosphere, little precipitation and partly absence of any tree-growth.

The northern part, known as the plains (54°—38° north latitude), is divided from north to south by the Rocky Mountains, and has on both sides the same flora, crossing the lower parts of the mountain-range. Amongst the characteristic plants are the most common several species of Artemesia, known under the name "Sage."

The southern part, comprising Arizona, southeast California, the southern parts of Nevada and Utah, New Mexico, West Texas and North Mexico is characterized by many species of Cactaceæ, and thorny, woody plants, mostly small trees. There are many peculiar Compositæ of which, out of two hundred and twenty-one North American genera, one hundred and sixty-five are represented in this province; and of these again seventy-one do not extend to California and to the East neither. Out of fifteen hundred and thirty-five North American species of this order seven hundred and forty-four we find there, and of those five hundred and forty-two only in this province, although many of them in tropic America.

2

5. South Florida is included, by Grisebach, in his large province of North American forests. In his "Geographical Distribution of West Indian Plants," he says that only a few woody plants, common to West India, occur in Florida and the Keys (fourteen he mentions). The reason why, I suppose, that the flora of South Florida is a part rather of the West Indian than of the North American flora, I have shown in an article in the American Naturalist about ten years ago. At that time, I proved that two hundred and forty-seven species, mostly West Indian, occur in South Florida that do not extend into Florida, or any of the South-eastern States; further, that amongst these there are not less than one hundred and thirty-six woody plants, and it must be remarked that a number of introduced species was not included. In the meantime a supplement to the Flora of Chapman was published, furnishing to the above number an addition of more than a hundred species of vascular plants, so that now three hundred and sixty species (mostly West Indian) are known in South Florida, that do not pass beyond the 29° of north latitude; one hundred and sixty-nine species belong to one hundred and thirty-four genera, which are not found farther northward, and of these again two hundred and ninety-five species are of sixteen orders, not represented in Northern Florida. Besides these we find in Chapman's Flora one hundred and eighty-nine species that do not pass over the northern State line of Florida, although some of them occur westward along the gulf shore. These are certainly of southern origin; so we have five hundred and fifty species, nearly all West Indian.

About fourteen hundred and forty species we find in Chapman's Flora, the habitat of which is said to extend to Florida, that means always the northern part. There is no full list of the plants of Southern Florida published, so that we could know how many of those northern species reach South Florida; but it is not probable that more than two-fifth of the above number will be found there. That would be equal to the number of southern species. As the exploration of the inner parts of the country, specially the everglades, is not finished, many more southern species may be found, when we consider that in the last ten years the number of known South Floridan species increased at the rate of forty-six per cent.

So we may conclude that the Flora of South Florida is composed of an equal number of North American and West Indian species, and it is probable that the latter will prevail.

There is no doubt that the fauna, especially the insect-fauna of a country, must be adapted to the flora; the areas of both will nearly be coincident, as we find it on a map representing the zoogeographical provinces in the third Report of the United States Entomological Commission (1883). And there we notice a fact of great interest: The southern part of Florida is separated from the Atlantic and united with the West Indian Province.

TOPOGRAPHY.

GEOLOGICAL FORMATION AND SOIL.

The country which the Illinois river is traversing is a plain, cut in by valleys 30 — 60 meter deep, and very slightly sloping from the northeast to southwest, as will show the elevations of the following points: The Mississippi, at low water at Dubuque, is said to be 186 meter; at the mouth of the Ohio, 83 meter; the water-shed between Lake Michigan and the Des Plaines river about 190 meter. Between Rock river and the Mississippi there are single hills rising to an elevation of 375 meter above the sea level, and 90 meter above the surrounding country. The River Des Plaines, running in a distance of a few miles along Lake Michigan, unites under 41°20' north latitude with the Kankakee coming from Indiana, and from there the river is called Illinois, running for sixty miles due west; then for about one hundred and eighty miles in a southwesterly direction, and empties into the Mississippi under 30°50' north latitude, 122 meter above sea-level. The descent is, in average, 14 decimeter per mile. About one hundred and forty miles above the mouth is Peoria, situated on the right bank upon two terraces, the first of which is 15 meter, the second, a little over 60 meter above low-water mark. Between the two terraces runs parallel with the river a depression, no doubt once a slough; the lower terrace, being an old sand-bank, rises at the lower end in a sand-hill, which was probably formed by a counter-current coming from the Kickapoo Valley, and shutting up the slough. This process was probably going on about the end of the drift period, but may be observed to-day on many rivers of the West. The second terrace, of equal height with the bluffs on the east side of the valley and about three miles distant, formed the oldest banks. The bluffs do not run in a continuous, straight line, but are interrupted by shallow or often deep ravines. The river, widening fourteen miles above Peoria, forms a sheet of water called Peoria lake, which at the lower end is about 1,600 meter wide. From there the stream keeps an average width of 270 meter.

In spring the river does rise from low water (140 meter above sea-level), 6 meter and then on the left bank the bottom-land is overflowed. The lower end of the lake was, thirty years ago, much wider than now, a little creek coming from the east, often changing channel, formed in the meantime not less than eighty or one hundred acres of land, partly covered already with cottonwood and willows, and increasing in a direction against a narrow strip separating a slough from the river. Should, in the course of time, the opening, left yet, be shut up, the slough will, drying up, turn in a prairie: the same process that was going on in the past on the right bank. Two miles farther downward the river bends and the bluffs border the left bank, the bottom-landbeing on the right side, where the Kikapoo creek enters from the west through a narrow valley with steep bluffs and numerous coal mines.

The great coal-field, occupying two-thirds of the State of Illinois, is covered by the northern drift, which is spread over the Northern Mississippi Valley southward to 39° north latitude. That this drift once filled the whole river valley, and that this was afterward washed out, is sufficiently proved by the large bolders found along the river banks, granit, syenit, diorit, porphyr, etc., that were brought by icebergs from the far north and dropped during melting; they were left when the softer material was floated away.

The bottom soil is alluvial; on the upland we find, below the humus, a subsoil of alternating layers of loam and gravel. There is no limestone soil, or very rare. There is little chance in the flora of Peoria to make observations on relations between chemical qualities of the soil and vegetation, if we would not attribute significance to the growth of certain grasses on pure sand soil. It is always the physical condition the habitat of each species depends upon: exposition to sunshine or shade, loose or compact ground, dryness of humidity of the soil and similar contrasts are favorable or exclusive.

CLIMATE.

As generally in the middle parts of the great continents of the northern hemisphere our climate is an excessive one. Hot summers, cold winters and a rapid change of temperature at all seasons is the character of this climate.

TABLE 1.

TEMPERATURE, DEC. 1, 1855 TO NOV. 30, 1885.

	Before sunrise	7 a.m.	2 p.m.	9 p.m.	Mean	Max.	Min.	Frost Days.	Days Maximum not above freezing point.
December........	24.5	24.5	33.3	28.6	28.6	71	—22	24	13
January	19.4	19.4	29.5	24.3	24.3	65	—27	27	17
February	23.5	23.5	35	29.8	29.2	70	—15	22	10
March	31	32.4	44.8	37.6	38.1	79	— 6	18	5
April............	42.7	45.8	59.7	50.6	51.7	88	18	5	....
May	54.4	58.5	72.5	62.2	63.8	98	30	0.2	....
June............	63.5	68.5	81.6	71.6	73.3	100	35	......	....
July............	67.6	72.6	86.3	76.1	77.7	104	48	......	....
August..........	65.5	69.5	84.1	73.6	75.2	105	41	......	....
September	57.7	60.6	75.9	65.2	66.6	98	34	......	....
October..........	45.8	47.3	62.4	52.7	53.7	90	14	3	....
November	33.8	34.2	45.5	38.9	39.4	77	-- 1	14	3
Year........	44	46.4	59.2	51	51.9	105	—27	113	48

The range of the thermometer scale in thirty years comprised 132° and nearly the same (127) even in one year 1872, when in January the minimum was —22 and the maximum in August 105. The lowest stand of the mercury—27 was observed on the 5th of January, 1884. The greatest range in one month (January, 1874,) was 87° from —22 to 65, and the greatest range in twenty-four hours was observed 1876 in January 28th, 2 p. m., to 29th in the morning the mercury fell 53° from 61 to 8, and again 1881 in January 13th to 14th falling from 34 to—10. Such high daily oscillations are frequent, particularly in February, December and April, and even in July the greatest difference in twenty-four hours was 37° (1860 2d to 3rd.)

To show the march of the mean temperature from day to day during the year the mean for each day (in thirty years) is computed in table two.

TABLE 2.

DAILY MEAN TEMPERATURE, 1855 TO 1885.

	Dec.	Jan.	Feb.	March	April	May	June	July	Aug.	Sept.	Oct.	Nov.
1	34	23.1	24.7	35.9	44.5	55.9	68.4	75.4	77.9	71.4	61.7	46.6
2	34.8	23.2	24.7	35.4	46.5	55.4	69.5	76.5	76.9	71.4	63.2	45.5
3	34.1	24.1	24.3	32.1	48.5	58.6	69.9	77.7	76.2	71.7	62.5	45.2
4	34.7	22	23.1	30.7	48	60.5	68.8	78.3	76.3	71.5	59.3	47.6
5	34	23.1	25.8	34.4	47.8	60.2	70.8	78.6	76.9	71.4	58.6	44.7
6	32.7	22.6	29.5	36.9	48.2	59.3	70.6	78.8	76.9	70.9	58	45.6
7	30	22	28.4	35.4	49.1	62.1	71.9	78.3	77.6	70.9	60	47.3
8	28	20.7	27.4	36.8	47.9	63.3	70.9	79.2	77.4	70.5	58.5	45.9
9	30	21.6	24.6	35.5	48.5	62.1	70	79.1	77.3	69.7	58.3	43.2
10	30	23.5	28	37.2	48.7	61.3	69.6	77.1	76.8	68.5	56.9	43
11	30.6	25.7	30.9	35.8	49.6	61.6	71.4	77.3	76.5	67.5	55.3	43.4
12	32.8	28.5	30.1	36.2	48.8	62	72.2	78.2	76.3	67.2	54.9	40.8
13	31.4	24.6	27.1	36.5	51.9	60.8	72.4	78	75.2	67.6	54.5	40.6
14	27.7	24.8	27.6	38.8	52.4	61.6	74.1	79.9	75.2	67.8	54.9	38.3
15	27.5	25	29.5	37.4	49.3	62.9	73.4	79.8	74.9	68.2	55.2	38.4
16	28.4	22.1	30.2	35.7	49.5	63.5	73.4	80.4	75	67.8	54.8	40.2
17	26.6	21.1	29	36.3	52.6	64.2	72.9	77.6	75.4	66.4	52.9	38.9
18	27.6	23.4	30.2	37.9	54.6	64.9	73.5	77.5	76.5	65.9	50.2	37.1
19	27.3	24.7	29.7	38.4	54.7	66	73.5	77.2	75.8	64.3	50.7	35
20	25.7	25.4	30.9	36.9	53.9	65.9	73.7	76.9	75.4	61.5	51.5	35.1
21	26.1	25	30.1	37.1	54.7	64.6	73.9	76.4	74.9	62.3	52	34.8
22	22.8	25.2	32.3	41.4	54.4	66.5	76.6	76.6	73.8	64	49.3	35.2
23	21.9	25.4	30.5	40.4	52.5	68.2	78.4	77.1	73.9	65.6	47.9	32.8
24	24.7	25	31.3	39.5	53.8	70.3	78.4	77.9	73.8	63.9	49.4	33.1
25	26.2	26.1	33.8	39.6	56.1	68.8	76.8	78.4	72.6	62.8	50.4	35.6
26	27.4	28.3	35.2	42.3	57.2	68.2	77.6	77.5	73.4	62.6	49.6	35.7
27	26.2	27.1	34.7	43	56.6	67.4	77.1	78	74.4	63.3	49.3	34.6
28	27.5	25	34.6	43.9	55.6	67.7	77.2	77.1	72.6	62.6	49.9	33.9
29	25.9	25.3	......	42.9	57 6	68.1	76.7	75.9	71.3	62	47.8	31.6
30	25.8	27	......	44.5	56.2	67.9	76.7	77.3	71.9	60.8	45.6	31.8
31	24.4	25.7	......	45.6	......	69	......	74.5	73	......	42.7	......

From the 5th of March the mean keeps above the freezing point, rising by and by with little oscillations to the maximum mean on the 16th of July (80.4,) then it falls slowly to the middle of September, and more rapidly to the 29th of November, when it reaches the freezing point, under which it keeps from the 7th of December to the 21st of February, with the only exception on the 12th of December. From the 22d of February to the 5th of March only the 23d and 24th of February and the 4th of March have the mean temperature below freezing point.

There are small oscillations in spring and fall, the depressions of which may be filled after longer observations, but those in winter are too great, not to be considered as more constant, as well as those between the latter part of May and the end of July.

Comparing the result of the thirty years observations with the march of temperature of a single year we find the latter much more irregular. In 1858 after a very mild January followed a severe February, and the first decade of March was colder than the first decade of January. In 1869 the first decade of February was the coldest of that winter and March colder than January.

The mean temperature of Peoria is about the same as that of Paris, in France, under 48° 50′ north latitude, that is 8° farther north than Peoria. We find for spring and fall about the same temperature in both localities, but a great difference for summer and winter.

	WINTER.	SPRING.	SUMMER.	FALL.
Peoria	27.4	51.2	75.5	53.2
Paris	38.5	50.5	64.6	52.5
Difference ...	8.9	0.7	10.9	0.7

Rome, in Italy, about 1° farther north than Peoria has a mean temperature of 60.8, in summer 74.3, in winter 50. That makes the mean temperature of Rome about 9° and in winter 22.6° warmer than in Peoria and the summer is nearly 1° colder. These examples may be sufficient to show the difference of climate of western Europe and the central parts of North America.

WINTER.

TABLE 3.

	FROST DAYS.		Days betw'n	No. of Frost Dayr.	Days not above freezing point.	FROST DAYS.		
	First.	Last.				Before.	Dec. Jan. Feb.	After.
1856-7	Oct. 1	May 11	223	142	63	26	74	42
1857-8	Oct. 20	Apr. 26	189	112	36	19	73	20
1858-9	Oct. 9	Apr. 23	197	96	35	14	67	15
1859-60	Oct. 9	May 1	206	120	47	17	82	21
1860-1	Oct. 12	Apr. 19	190	116	53	20	74	22
1861-2	Oct. 24	Apr. 6	165	114	57	17	78	19
1862-3	Oct. 24	Apr. 8	167	112	22	28	58	26
1863-4	Oct. 6	Apr. 21	198	109	38	23	66	20
1864-5	Oct. 9	Apr. 24	198	104	44	16	72	16
1865-6	Oct. 28	Apr. 7	162	107	50	12	75	20
1866-7	Oct. 31	May 8	190	118	61	12	73	33
1867-8	Oct. 24	Apr. 18	178	115	58	14	84	17
1868-9	Oct. 8	Apr. 14	189	112	45	14	70	28
1869-70	Oct. 13	Apr. 17	187	126	41	33	73	20
1870-1	Oct. 31	Apr. 1	153	92	38	10	71	11
1871-2	Oct. 28	Apr. 22	178	130	57	16	85	29
1872-3	Oct. 11	Apr. 25	197	124	73	20	84	20
1873-4	Oct. 7	Apr. 29	205	130	40	29	68	33
1874-5	Oct. 12	May 2	203	131	74	20	80	31
1875-6	Oct. 11	Apr. 5	178	98	36	19	55	24
1876-7	Oct. 15	Apr. 5	173	122	61	16	79	27
1877-8	Nov. 3	Mch. 25	143	51	15	9	41	1
1878-9	Oct. 19	Apr. 5	169	113	53	14	80	19
1879-80	Oct. 24	Apr. 12	172	92	28	16	54	22
1880-1	Oct. 18	Apr. 16	181	143	83	26	84	33
1881-2	Nov. 9	Apr. 12	155	93	24	15	61	17
1882-3	Nov, 12	Apr. 24	164	120	63	13	80	27
1883-4	Nov. 1	Apr. 8	160	106	58	13	76	17
1884-5	Oct. 23	Apr. 13	173	112	67	12	74	26
Mean	Oct. 18	Apr. 17	180	112	48	18	72	22

The three winter months together had the lowest mean 20.7 in the winters from 1872 to 1873 and 1874 to 1875.

Above freezing point was the mean of the winters 1862—63, 1875—76, 1877—78, and 1879—80; in all the rest it was below freezing point.

The coldest January was that of 1857, 13.5; the coldest February 1875, 15.5; the coldest December 1876, 18.5.

The warmest January was in 1880, 40.9; the warmest February 1878, 37.5; the warmest December 1877, 44.3.

The coldest decade in January was 1864 1st and 10th, 0.2; in February, 1875, 10th to 20th, 8, and in December, 1872, 21st to 31st, 8.8.

The warmest decade in January was in 1864, 21st to 31st, 41.8; in February, 1871, 21st to 28th, 41.2; in December, 1862, 21st to 21st, 41.7.

When we call the three months December, January and February the three winter months, it is obvious that this is mere theory. Practically winter is not restricted to those three months; there are no general limits which are good for every year. When we take freezing as a distinctive quality of winter we find its limits very variable in different years. The mercury is falling below freezing point in a period commencing on the 1st of October and ending on the 11th day of May, so that the first frost days in the thirty years occurred between the 1st of October and the 12th of November; the last between the 25th of March and 11th of May.

The longest of those periods was in the winter from 1856 to 1857, the first frost was noticed on the 1st of October and the last on the 11th of May, a period of 223 days.

The shortest was that from the 3d of November, 1877, to the 25th of March, 1878, a period of 143 days. The former contained 142, the latter only 51 frost days. Computing the average we find the first frost day to be the 17th of October, for the last frost day the 17th of April, a period of 183 days with 112 frost days, and 48 days with a mean temperature not rising above freezing point.

SPRING.

The mean temperature of the three spring months together is 50.2. The lowest mean was observed in 1857 43; the highest in 1878= 56.6.

The coolest March, 1867= 29.5; the coolest April, 1857= 39.9; the coolest May, 1867=55; the warmest March, 1878 50.5; the warmest April, 1878 57.9; the warmest May, 1881 71.4.

The mean temperature of the decades are in March 1st=35; 2d= 37; 3d- =41.8; the lowest was the 1st in 1857 22.3; the highest the 1st in 1878 55.5. April 1, 47.8, 2d, 51.7, 3d, 55.5; the lowest the 1st= =32.7; the highest the 3d, 1879=66.2; in May the 1st =59.9, 2d= 63.3, 3d==67.9; the lowest the 1st in 1867 51.1; the highest the 3d in 1881, 77.

The highest mean temperature of a single day of March was in 1875, on the 30th=65.6, of April in 1872, on the 29th=77, of May in 1860, on the 24==85.1; the lowest of March in 1867 on the 13th=4.7, of April in 1857 on the 6th 23.7, of May in 1875 on the 1st 39.5.

The highest stand of the thermometer was observed in March, 1860, on the 30th= 79; April 1865, on the 26th- 88.5; in May 1860, on the 24th - 88.5. The lowest in March, 1867, on the 14th _6; in April 1887, on the 15th 18; in May 1867, on the 8th 30.

There are, in average, 18 frost days in March; 5 in April, and, in May, 5 were observed in 30 years. The most frost days we had were in March 1859 29; and in April 1857 18. There was no frost day observed in April 1878; only 13 p. c. of the frost days of April occurred after the 17th, at which date, for the last time, a mean temperature below freezing point was observed.

SUMMER.

The mean temperature of the three summer months is 75.5.

The coolest summer was in 1866 and 1869 73; the warmest in 1874 = 79. The coolest June in 1869 69; the coolest July 1865- =71; the coolest August 1866=70; the warmest June in 1873=79; the warmest July 1868=82.7; the warmest August 1881=80.5.

The mean temperature of the decades was of June 1st decade=70; 2d=73; 3d=77; of July 1st=77.9; 2d=78.3; 3d=77.2; of August 1st= 77; 2d =75.6; 3d=73.3.

The coolest decade of June was the 1st of 1863=63; July the 2d in 1865= 65.3; of August the 3d in 1863—65.3; the warmest in June the 3d in 1858=85; in July the 2d in 1878=89.8; in August the 1st in 1861=86.7.

Of the single observations the highest for June was on the 24th, in 1856=100; for July on the 15th, 1859, on the 4th, 1874, and on the 30th, 1885=104; for August on the 31st in 1873=105; the lowest for June on the 4th 1859—35; for July on the 2d, in 1861, and the 16th, in 1863=50; for August on the 29th, in 1863=41.

FALL.

The mean temperature of the three fall months is 53.3.

The coolest fall was in 1880 =48.9; the warmest in 1884= 55.1; the coolest September was in 1866=60.5; the warmest in 1865=73.1; the coolest October in 1869=48.2; the warmest in 1879=62.7; the coolest November in 1880=30.2; the warmest in 1867=44.4.

The mean temperature of the decades are the following: For September 1st 70.8; 2d= 66.4; 3d=63. For October 1st=59.7; 2d 53.5; 3d 48.5. For November 1st=45.5; 2d=38.8; 3d=33.9.

The coolest decade in September was the 3d in 1856=52.7. The warmest the 1st in 1884—81.3. The coolest in October the 3d in 1869= 36.3. The warmest the 1st in 1879 76.3. The coolest in November the 3d in 1880=20.3. The warmest the 1st in 1874 -54.5.

The highest stand of the thermometer was observed on the 3d of September, 1864, and on the 5th of September, 1881 98; on the 3d of October, 1856, on the 12th of October, 1879, and on the 8th of October, 1884 90; on the 7th of November, 1874 77. The lowest on the 29th of September, 1871=34; on the 24th of October, 1869 =14, and on the 23d of November, 1857=_15.

By comparison of the temperatures of different places in Illinois during the meteorogical year, December 1869 to November 1870, we find in the mean temperature of Peoria and Springfield which is nearly a degree farther south, and Ottawa which is more than half a degree farther north, scarcely any difference, but Galesburg farther west and on a higher elevation had, that same year, a mean of one degree lower and a January very much colder. Of the same year the temperatures of Steubenville, Ohio, Fort Madison on the Mississippi, and Nebraska City on the Missouri, all nearly in the same latitude with Peoria, as compared, show the following figures:

3

	MEAN OF THE YEAR.	IN WINTER.	IN SUMMER.
Steubenville	54.5	34.3	75.2
Peoria	54.	29.6	76.6
Ft. Madison	52.3	27.8	76.6
Nebraska City	52.	27.5	74.8

The means of the year are decreasing from east to west; in the same way lower the temperatures of the winter, but the summer is the hottest on the Mississippi and on the Illinois, well considered that Steubenville and Nebraska City be on a greater elevation above the sea-level, and that the climate of Steubenville is influenced by the Canadian lakes.

By a mean period of frost of one hundred and eighty-three days for the season free of frost one hundred and eighty-two days would be left, and so the year would equally be divided; but as the last frost day in thirty years occurred on the 11th of May and the first on the 1st of October, there would be left only one hundred and forty-two days, and even that is good only for the locality of the observations in the midst of the city, for on exposed places in the open country, even in this period, frosts may occur, and, indeed, on the 4th of June, 1859, when the thermometer in the city showed a minimum of 35°, and on the 29th of August, 1863, when the mercury went down to 41°, frosts were reported from the sur·rounding country. Moreover, the so-called "white frost" may be formed at a temperature of the air above freezing point. All bodies radiate heat, and their temperature lowers, when they do not receive a fresh supply of heat from outside. So do the plants at night time. Radiation takes place in all directions to the surrounding air, and the more so the more clear the sky is and the more calm the air. A small thermometer placed in the grass, on an unprotected place, may very likely show ten or more degrees less than that one that is suspended five feet above the ground. The plants exhale constantly water in gas form, which precipitates upon the cooled surface, and when that cooling reaches the freezing point, white frost is formed.

The difference of temperatures observed in localities of the same latitude shows, that meteorological observations of one locality are good only for that locality, and perhaps its next vicinity, and it is lost labor to compute averages for wider districts f. i. of the State of Illinois, divided by straight lines in a northern, central and southern part, or for even larger area of five or six States, comparing the results with the crops of the same districts so different, not only of temperature and precipitation, but in the nature of the soil. There is no more sense in it than would be in computing the temperature of the whole of North America. It is only waste of time and paper.

The means of the single years range between 8°, the lowest mean temperature of a year was that of 1857 =48.7; the highest that of 1878 =

56.7. The mean of the first ten years was 52.1, of the second, 51.4, of the last, 52.7.

TEMPERATURE ACTING UPON PLANTS.

One plant is more sensible than others; cultivated plants introduced from a warmer climate more than indigenous ones, and often in fall the leaves of the tomatoes and dahlias may be killed by a temperature of the atmosphere, which is above freezing point. On the other hand I observed in April 1857, when the gooseberry bushes were green for a week, that the mercury descended to 19° without injury to the leaves.

In Transact. of Ill. Agric. Soc., Vol. III, is published a paper read - before the Illinois Natural History Society in June, 1859, "On Meteorology in connection with botanical observations," in which I did show that each plant require a certain sum of heat in a certain space of time to perform its physiological functions, and that the degrees below the freezing point, if not destructive, be not reactive, but inactive, and, therefore, all the degrees below freezing point be of no account and excluded from the computation.

In the summer of 1857 I made some observations on the growth of Indian corn. On the 16th of May, two days after a heavy rain, I planted some corn in the yard of my residence, it sprouted on the 25th of May and was ripe on the 30th of September. During these 138 days the sum of daily mean temperature, 5 feet above the ground in the shade, was 3064 c.

The sum of daily mean temperature of the soil four inches below the surface, at 3 P. M., was 3,443; the quantity of rain, 13.2 inches; the mean humidity of the atmosphere, 68 p. c. of saturation. The result of this observation is about the same as that reported by Bousingault upon an observation made at Alais in South France (44° N. L.)

In the above observation, during the period of which the minimum of the temperature was not falling below freezing point, the meteorological observations as made for the Smithsonian Institution, and now for the Signal office, could be used, and the means were computed from the three daily observations at 7 A. M., 2 P. M. and 9 P. M., although this mode of calculation does exclude for the whole summer the minimum (before sunrise) and the maximum (about 3 P. M.); and so the above sum, which would be necessary to ripen Indian corn, was obtained, provided that the temperature of each degree above freezing point have any effect upon its growth.

That, in winter, the temperatures below freezing point are not reactive, can be proved by observations of the periods of blooming of woody plants.

In 1857 spring was tardy, in 1859, very early. Comparing the time of flowering of certain species with the sum of daily mean temperature, commencing with January and excluding all temperatures below freezing point, it is surprising to see the coincidence of figures, and the great difference when the negatives are not excluded. A table published in the above named transactions, but full of printing errors, and, therefore, herewith corrected, will prove that.

First Day of Blooming.	Negatives not excluded	SUM OF DAILY MEAN TEMPERATURE (CENTIGRADE), NEGATIVES EXCLUDED.		
		Above freezing.	1 degree above freezing.	2 degrees above freezing.
Acer Saccharinum { May 10, 1857	39.4	526 on 87 days	443 on 86 days	392 on 85 days
Apr. 20, 1859	327.4	522 „ 88 „	447 „ 87 „	377 „ 84 „
Crataegus { May 20, 1857	155.6	642 „ 97 „	549 „ 96 „	489 „ 95 „
Subvillosus { Apr. 30, 1859	452.4	646 „ 98 „	562 „ 97 „	483 „ 97 „
Aesculus { May 20, 1857	155.6	642 „ 97 „	549 „ 96 „	489 „ 95 „
Glabra { Apr. 30, 1859	452.4	646 „ 98 „	562 „ 97 „	483 „ 97 „
Cerasus { May 25, 1857	263	750 „ 102 „	652 „ 101 „	587 „ 100 „
Virginiana { May 5, 1859	555	750 „ 103 „	661 „ 102 „	576 „ 102 „
Asimina { May 31, 1857	361.2	848 „ 108 „	744 „ 107 „	672 „ 106 „
Tribola { May 10, 1859	657.5	852 „ 108 „	758 „ 107 „	668 „ 107 „
Robinia { Jun. 17, 1857	471	978 „ 115 „	867 „ 114 „	788 „ 113 „
Pseudacacia { May 16, 1859	767	962 „ 114 „	862 „ 113 „	766 „ 113 „

Eleven series were calculated in that way as far as ten degrees above freezing point. As the figures diverge more and more, it seems that the sap of our woody plants moves as soon as the temperature rises above freezing point, for there the figures come nearer together. Only Robinia makes an exception, the starting point of which is probably one degree above freezing point.

This is a Southern tree and at Peoria introduced, but as the observations were made on an individual, standing right near the place of observations, these are the most reliable.

How much later these plants would have been in bloom in the year 1857, when the negative temperatures acted reactive instead of inactive, can be proved by the following table:

	Jan.	Feb.	Mar.	Apr.	May	Jan.—May.	Feb.—May.
1857.							
Sum of daily mean temperature	—320	47	27	142	465	361	681
Sum of daily mean above freezing	6	106	108	163	465	848	842
Number of days with a temperature above freezing	4	24	21	28	31	108	104
1859.							
Sum of daily mean temperature	—73	26	225	275	615	1068	1141
Sum of daily mean above freezing	57	84	226	279	615	1261	1204
Number of days with a temperature above freezing	19	18	31	30	31	129	110

It may be doubted whether it be advisable to include the temperatures of January. In continuously cold winters a temperature rising only a few day a few degrees above freezing point, may be inactive; but we have, not rarely, quite warm winter months in which the buds of the trees considerably swell, and, when cold weather follows, rest stationary for a time. The time of blooming of a number of woody plants were noted, amongst which Amelanchier canadensis. From seventeen years an average sum of heat of 450 degrees (centigrade) was found from the 1st of January to the day of blooming, which is in average the 21st of April. In this period of one hundred and eleven days the temperature rises on seventy-eight days above freezing point. The earliest time of blooming was observed in 1871, on the 4th of April, with a sum of 414 degrees; the latest in 1857, on the 8th of May, with a sum of 496 degrees. The difference of 82 degrees may be accounted for, when we consider that the heat is not the only agent. The time of blooming may partly depend on the moisture in the ground, the dryness of the atmosphere, and, before all, on the quantity of light and direct insolation.

The above figures indicate not the absolute, but the relative value of heat, i. e.: When the sum of daily means above freezing point reaches 450 degrees in the shade, then the Amelanchier is in flower, or is in flower since several days, when we had much sunshine and the air is dry, or will be in flower in a few days, when we had less sunshine and the air is moist.

Of fifteen other species, the necessary heat and the time of blooming was calculated.

	MEANTIME OF BLOOMING.	SUM OF HEAT IN CENTIGRADE DEGREES.	NUMBER OF OBSERVATIONS.
Acer dasycarpum...................	March 27th.	210	11
Ulmus americana	March 31st.	230	13
Negundo aceroides	April 21st.	440	7
Acer saccharinum	April 28th.	530	12
Prunus americana..................	April 29th.	550	6
Cercis canadensis..................	May 3d.	600	12
Aesculus glabra	May 6th.	650	15
Pyrus coronaria....................	May 11th.	740	5
Morus rubra	May 13th.	760	5
Prunus virginiana.................	May 13th.	770	7
Asimina triloba	May 15th.	800	7
Prunus serotina	May 22d.	920	5
Robinia pseudacacia...............	May 23d.	940	15
Catalpa speciosa...................	June 7th.	1270	6
Tilia americana	June 26th.	1700	11

That the temperature of December is not of great influence, as one might suppose, shows the December, 1877, with a sum of 225 degrees above freezing point. The American Elm was in bloom on the 8th of March, 1878, with a sum of 235 starting from the 1st of January, and the same on the 10th April, 1866, with a sum of 240. The December, 1865, had only a sum of 40 degrees. So the elm must either be in bloom on the 4th of February, when the sum of heat was 280 degrees, or required, in 1878, a sum of 460 against 280 in 1886. Amelanchier was in bloom in 1878 on the 27th of March, with a sum 454 degrees, and in 1861, on the 11th of April, with 453 degrees. December, 1860, had only a sum of 27 degrees; accordingly, Amelanchier required 679 degrees in 1878, and only 480 in 1861, or the due time of blooming was in 1878, the 8th of March.

BAROMETER.

The observations on the pressure of the atmosphere comprise twenty-five years from December, 1860, to November, 1885.

The mean reduced to freezing point was 29.628 inches; the mean, at 7 A. M., is 29.644; at 2 P. M., 29.606; at 9 P. M., 29.634. The highest stand was observed in January, 1866, 30,671, and the minimum in April, 1880, 28,581, the range being 2.090.

The greatest range in one month was observed in January, 1866, 1.676; the smallest in August, 1878, 0.283. The highest mean of a month had, December, 29.698; the lowest May, 29.548.

The greatest range in twenty-four hours was observed in January, 1.028. In July it is only 0.389.

There are generally two oscillations in twenty-four hours, with two minima at 11 A. M. and 10 P. M., and two maxima at 4 A. M. and 4 P. M. The rise and falling is, in the tropic countries so regular, that it is possible to determine the daytime from the stand of the barometer; in our zone it is more variable, so that often a continuous falling or rising for several days is observed.

TABLE 4.

BAROMETER REDUCED TO FREEZING POINT.

DECEMBER 1860—1885.

	7 A.M.	2 P.M.	9 P.M.	Mean.	Maximum.	Minimum.	Range.	Greatest change in 24 hours.
December	29.714	29.680	29.717	29.704	30.389	28.771	1.618	1.017
January	29.724	29.685	29.726	29.712	30.671	28.795	1.876	1.028
February	29.688	29.649	29.683	29.673	30.453	28.823	1.630	0.967
March	29.632	29.600	29.632	29.621	30.364	28.612	1.752	0.950
April	29.572	29.540	29.568	27.560	30.252	28.581	1.671	0.811
May	29.571	29.532	29.554	29.552	30.041	28.670	1.371	0.750
June	29.583	29.546	29.565	29.565	29.956	28.996	0.960	0.508
July	29.607	29.572	29.587	29.589	29.944	29.134	0.810	0.484
August	29.608	29.572	29.590	29.591	29.965	29.207	0.758	0.492
September	29.670	29.626	29.651	29.648	30.083	29.051	1.031	0.617
October	29.677	29.629	29.664	29.657	30.254	29.025	1.229	0.718
November	29.679	29.637	20.678	29.664	30.308	28.725	1.583	0.838
Year	29.644	29.606	29.634	29.628	30.671	28.581	2.090	1.028

PRECIPITATION.

The mean quantity of rain and melted snow was 35.6 inches per year in one hundred rainy days. The smallest quantity falls in January, 1.6 in seven days; the greatest in June and July, each with four inches in ten and nine days The precipitation in winter is 6.1, in spring 9.7, in summer 11.2, in fall 8.6. This would be favorable when distributed in that way every year; but the single years differ very much. In 1856 it was only 22.8; in 1858, 51.4. There are sometimes long droughts. From the 29th of August to the 8th of October, 1871, there was only one rainy day in the middle of September with 0.65 of an inch of rain. The longest period without any rain was in 1861, in October and November, which lasted. twenty-eight days. There was one in the spring of 1863 of twenty-one days, in April and May, 1863, of twenty days; in July, 1873, of nineteen days, and the same in July and August, 1869. Sometimes there are long periods of too much rain, f. e. in 1858, from the 29th of April to the 10th of June, 15.7 inches in twenty-seven rainy days.

The quantity of rain is of less importance than the number of rainy days and their distribution. The highest number for one month was 18 in May, 1858, and in July, 1865, the lowest in September, 1871, and February, 1877, each with one rainy day.

Supposing that eleven inches of rain in twenty-six days of the three summer months be most beneficial, and that a plus or minus of two inches and two rainy days be of no importance, than we had in the summers of 1862, 1869 and 1872, a great excess in quantity, viz: 9.1, 7.8 and 10.8 inches surplus, and an excess in the number of rainy days in 1865 and 1866, viz: thirteen, and seven surplus. A deficiency in quantity show the years 1870, 1868 and 1865 with 6.6, 5.8 and 5.6 minus, and in rainy days, 1863 and 1856, viz: twelve, and eight minus.

The most normal summers (in regard of rain) were 1857 and 1871.

The greatest quantity of rain for one month was measured in May, 1858, 10.64; then in June, 1872, 9.76, and in September, 1875, 9.61.

The mean precipitation of the single months are: December, 2.5; January, 1.6; February, 2; March, 2.7; April, 3.2; May, 3.8; June, 4; July, 4; August, 3.2; September, 3.5; October, 2.7 and November, 2.4.

HUMIDITY OF THE ATMOSPHERE.

The relative humidity of the air was computed from the difference of the wet and dry thermometer by means of Guyot's tables. When there is no difference the atmosphere is saturated with moisture, and that is noted by 100. The greater the difference the lower is the per centage; 20 means very dry, and there is scarcely ever noted a lower figure.

The mean of the year is at 7 A. M., 81; at 2 P. M., 58; at 9 P. M., 75. The highest mean in January at 7 A. M., is 89; the lowest in May, 2 P. M., 50.

The pressure of vapor is the highest in July, 9 P. M., 0.669 of an inch, the lowest in January, 7 A. M., 0.114.

The means for the year are: 0.316 at 7 A. M.; 0.338 at 2 P. M., and 0.340 at 9 P. M.

TABLE 5.

PRECIPITATION, HUMIDITY OF ATMOSPHERE, AND

PRESSURE OF VAPOR.

	MEAN PRECIPITATION		RELATIVE HUMIDITY OF THE ATMOSPHERE.			PRESSURE OF VAPOR IN FRACTIONS OF INCHES.		
	In Inches	In days	7 A.M.	2 P.M.	9 P.M.	7 A.M.	2 P.M.	9 P.M.
December...............	2.48	8	87	70	80	.130	.142	.134
January	1.61	7	89	71	81	.114	.130	.122
February	2.04	8	87	66	80	.122	.145	.138
March...................	2.68	9	80	59	74	.157	.185	.177
April....................	3.17	10	75	52	70	.240	.260	.263
May.....................	3.80	10	73	50	70	.362	.390	.397
June	4.03	10	77	53	74	.537	.559	.567
July	4.04	9	78	53	74	.626	.642	.669
August	3.16	7	81	52	75	.590	.610	.642
September..............	3.49	8	83	55	76	.465	.500	.492
October	2.71	7	83	55	75	.276	.299	.291
November	2.37	7	81	63	75	.173	.193	.185
Winter	6.13	23	88	69	80	.122	.139	.131
Spring..................	9.65	29	76	54	71	.253	.278	.279
Summer	11.23	26	79	53	74	.548	.604	.626
Fall	8.57	23	82	58	75	.305	.331	.323
Year	35.58	100	81	58	75	.316	.338	.340

CLOUDINESS AND SUNSHINE.

The cloudiness of the sky is expressed by figures, which mean the percentage of covering, 100 was noted, when the sky was entirely covered, 50 when half, and so on, and 0 when cloudless. The sky is the most covered in December in the morning, and least in August in the evening.

The mean for the year is 47; the highest for the month is in December, 55; the lowest August, 35.

From the amount of cloudliness cannot be deduced the time of sunshine during a period, for the sky may be half covered, the sun may shine during the whole day. It is necessary to note the time of sunshine every day. This was done from December 1857 to November 1868 and the result was that we had sunshine 58 p. c. of the time from sunrise to sunset.

The sunniest months are June and August each with 71 p. c.

How great the influence of insolation must be upon the growth of plants is shown by the difference of the thermometer in the shade and exposed to the sun, which, in June, often exceeds 20 degrees and more yet in winter.

	PER CENT. OF COVERING.				NUMBER OF DAYS				SUNSHINE.	
	7 A.M.	2 P.M.	9 P. M.	Mean.	Cloud-less.	Moder-ately cloudy.	Very cloudy.	With-out sun shine.	Hours.	per ct.
December	59	54	53	55	4	10	17	10	129	45
January	57	56	48	54	4	10	17	9	133	46
February	52	55	49	52	3	11	14	7	149	51
March	53	56	49	52	3	12	16	6	182	50
April	51	57	45	51	2	11	16	5	192	49
May	45	50	35	43	3	13	15	3	260	61
June	44	48	32	41	2	16	12	1	315	71
July	37	47	29	38	2	18	11	1	314	69
August	37	45	27	35	4	18	10	1	299	71
September	44	45	31	40	4	14	12	2	216	58
October.......	47	46	35	43	6	12	13	5	202	59
November	55	56	49	53	3	10	17	9	148	51
Winter	56	55	50	54	11	31	48	26	411	48
Spring	49	54	43	49	8	37	47	14	643	53
Summer	39	47	29	38	7	52	33	3	929	70
Fall...........	48	49	39	45	13	36	42	16	566	56
Year.........	48	51	40	47	39	156	170	59	2550	58

WIND.

West winds are prevalent from October to April. South winds during the summer; only in August east equals the south. About 12 p. m. of all the observations are marked as high winds, gales or hurricanes, but the force of winds were not measured by the anemometer but only estimated, and the dates are not quite reliable.

The windiest months are March and April; the calmest, August and September.

Wind and temperature, wind and cloudiness, wind and precipitation are, in a certain degree, correlative. The warmest winds are south, southwest and east; the coldest, northwest, north and northeast. The difference between the coldest and warmest winds is about 15, in spring even 20 degrees. Above the average is the temperature with south and southwest in all the months, with east only in spring and fall. Southeast wind is too scarce, so that no reliable mean could be abstracted. The temperature of north is always below. Northeast is only in November, December and January above, and that may be accounted for by the great quantity of cloudiness that always accompanies these winds, preventing radiation. Northwest has only, in August, a temperature above average. The region from which this wind come is naturally a cold one. only during the sum-

4

mer months excessive heat is accumulating, what has the above effect upon these winds. The same reason is good for the west during all the summer months; in the rest the west winds are cooler.

Northeast brings the most cloudiness, and west the least; the west is the only one that has a cloudiness considerably below average.

The relation of wind and precipitation must be considered in a double way. When we compute the direction of wind in 1,000 observations of precipitation, then we find that we have 258 times south wind; 174 times east; 159 times northeast; 105 times southwest; 95 times west; 84 northwest; 79 north, and 46 times southeast. But when we reduce the observations of precipitation to 1,000 of each wind direction, then we find for each wind the following per mille. of rain observations: northeast, 317; southeast, 153; southwest, 132; south, 126; northwest, 124; east, 111; north. 93; west, 46. That shows that northeast is the prevalent rain wind. But the single months differ. In summer southwest brings the most rain, and nearly all the thunder storms come from southwest or northwest. The average number in a year is twenty-eight. ·

TABLE 7.

DIRECTION OF WIND FROM December, 1855, TO November, 1885.

REDUCED TO 1000.

	West.	South West.	South	South East.	East.	North East.	North.	North West.
December	280	96	207	43	136	56	78	101
January	308	103	216	25	110	47	87	104
February	290	57	217	32	149	57	113	85
March..................	259	62	188	38	167	64	114	108
April..................	214	70	208	28	197	78	117	88
May..................	184	67	244	43	235	66	111	50
June	209	84	302	39	222	39	63	42
July..................	193	97	253	55	220	60	89	53
August	188	84	239	30	248	61	92	58
September..................	160	74	279	36	211	52	126	62
October	233	82	240	32	157	66	111	79
November..................	203	69	233	37	157	53	83	95
Winter..................	293	86	214	33	131	53	92	98
Spring..................	217	67	214	37	200	69	114	82
Summer	197	88	264	34	230	54	82	51
Fall	229	75	250	35	168	57	107	79
Year..................	233	79	235	35	183	58	99	78

TABLE 8.

NUMBER OF OBSERVATIONS OF STRONG WINDS MORE THAN TWENTY-FIVE MILES AN HOUR.

DECEMBER, 1855 TO NOVEMBER, 1885.

ONE PER CENT. OF ALL OBSERVATIONS (32,765).

	West.	South West.	South.	South East.	East.	North East.	North	North West.
December	14	3	.....	.....	3	.....	.....	2
January	14	2	.....	.....	1	.....	.....	2
February	26	5	5	1	3	.....	.....	4
March	25	10	9	1	3	8	1	2
April	23	13	9	.....	3	4	.....	3
May	12	7	4	2	.....	2	2	1
June	8	5	2	.....	.....	2	.....	.....
July	8	4	2	.....	.....	.....	1	4
August	.....	.....	2	.....	.....	.....	.....	1
September	3	1	6	.....	.....	.....	1	.....
October	11	8	8	1	.....	2	.....	2
November	17	2	6	3	.....	2	.....	1
Winter	54	10	5	1	7	1	.....	8
Spring	60	30	22	3	6	14	3	6
Summer	16	9	6	.....	.....	2	1	5
Fall	31	11	20	4	.....	4	1	11
Fall	161	60	54	8	13	20	5	22

TABLE 9.

MEAN TEMPERATURE COINCIDENT WITH CERTAIN WINDS.

DECEMBER, 1865 TO NOVEMBER, 1885.

	West.	South West.	South.	South East.	East.	North East.	North.	North West.
December	22.2	30.8	35.4	34.3	29.8	32.2	25.5	25.3
January	16.6	29.6	32.3	29.9	24.9	25	21.9	21.7
February	22.7	33.6	37.5	28.6	29.8	28.1	22.5	23.6
March	34.4	43.4	45.5	49.6	36.5	34.9	32.6	33.3
April	50.4	64	60.1	58.2	50.3	43.1	47.4	45.2
May	59.4	75.5	72.4	72.1	63.4	56.3	60	57.1
June	72.7	78.8	76.7	73.5	72.4	67	71	67
July	79.5	82.5	81.9	78.5	77	71.4	73.8	75.9
August	76.5	79.9	79.9	78	74.4	70.2	69.6	74.9
September	66	78.5	72.1	67.8	66.5	63.1	63.7	61.9
October	49.7	59.4	60.8	57.5	53.7	56	47.5	48
November	34.2	42.9	45.9	43.1	42.7	40.1	35.1	34.9
Winter	20.3	30.9	35	31.4	28.2	28.7	23.2	23.6
Spring	46.9	61.5	60.5	59	51.6	44.3	46.1	42.3
Summer	76.3	80.7	79.3	77	74.9	69.2	71.4	72.8
Fall	46.9	61.1	60	55.3	56.6	53	50.9	46.2
Year	45.4	60.1	59.5	55.8	55.8	46.6	47.3	42.2

TABLE 10.

WIND AND CLOUDINESS.

PERCENTAGE OF COVERING.

	West.	South West.	South.	South East.	East.	North West.	North.	North West.
December	35	50	61	72	72	88	61	70
January	31	48	59	66	71	90	63	71
February	33	54	55	53	61	93	61	56
March	31	52	57	60	60	87	56	62
April	38	48	51	51	52	81	52	56
May	32	51	42	55	40	70	44	60
June	30	55	38	45	40	61	42	55
July	30	45	37	35	37	56	37	37
August	28	42	35	36	30	65	39	45
September	30	42	37	33	35	78	38	51
October	31	36	42	60	37	80	43	65
November	38	49	54	66	44	90	55	72
Winter	33	50	58	64	68	90	62	67
Spring	33	50	50	55	50	71	51	60
Summer	29	47	37	39	35	61	38	46
Fall	34	42	44	52	44	82	44	64
Year	32	47	47	52	47	79	50	61

TABLE 11.

DIRECTION OF WIND IN ONE THOUSAND OBSERVATIONS

OF PRECIPITATION.

	West.	South West.	South.	South East.	East.	North East.	North.	North West.
December	81	89	231	61	190	170	78	100
January	37	66	288	18	188	162	100	140
February	71	55	219	32	178	216	139	90
March	84	84	228	49	162	192	106	95
April	98	75	191	44	228	191	98	75
May	85	115	272	51	203	141	82	51
June	110	186	318	56	161	79	31	59
July	147	179	244	38	172	103	55	62
August	106	159	237	21	163	155	65	94
September	111	88	306	24	182	138	84	67
October	93	100	270	60	117	194	83	83
November	107	55	287	87	142	180	42	100
Winter	65	71	244	39	185	183	104	109
Spring	89	92	231	48	197	174	95	74
Summer	121	177	272	39	166	107	48	70
Fall	104	81	288	56	147	170	70	84
Year	95	105	258	46	174	159	79	84

TABLE 12.

OBSERVATIONS OF PRECIPITATION REDUCED TO ONE
THOUSAND FOR EACH WIND.

	West.	South West.	South.	South East.	East.	North West.	North.	North West.
December	37	120	143	185	180	391	128	123
January..................	11	62	129	73	167	333	111	130
February	30	117	123	119	146	465	148	130
March	43	169	160	165	128	392	123	115
April..................	58	138	117	202	148	312	108	109
May	62	230	150	158	115	289	100	136
June	69	294	138	172	95	269	64	194
July	80	193	100	114	81	179	64	121
August	49	171	87	62	57	221	61	143
September	76	129	121	72	94	288	63	120
October	43	132	121	202	79	318	74	113
November..............	39	85	132	255	111	361	80	112
Winter....	26	95	132	136	164	398	130	126
Spring................	54	179	142	171	130	330	110	117
Summer...............	66	216	110	128	77	216	63	147
Fall	49	117	128	176	94	322	71	114
Year	46	132	126	153	111	317	93	124

VEGETATION.

The floral district of the upper Mississippi, as I understand it, will
comprise the southern part of Wisconsin from about the '44° N. L. Minnesota, from 46° N. L. southward, the greater southeastern part of Iowa
the wooded eastern borders of Nebraska and Kansas. Missouri, northwest-
ern Arkansas with the eastern borders of Indian Territory, and, finally,
Illinois — except the southern and southeastern borders along the Ohio
and lower Wabash (Ohio flora). There may be added the prairie land of
northwest Indiana, but the shores of Michigan Lake all around, should be
excluded as a part of the Canada flora. The western line of this district
would nearly coincide with the limits of the palæozoic formation on the
east, and the new red sandstone on the west side; but, as already said, it
is difficult to circumscribe a floral district by lines. Many western species
cross the Missouri into Iowa, and not a few the Mississippi into the Illinois
country and are found in the numerous and often extensive prairies between
the woodland. Therefore, the vegetation of the upper Mississippi should
be regarded a gradual transition between the woodland-flora of the Ohio
district and the prairie-flora of the western plains.

There are good reasons to exclude that part of the State of Illinois
that is south of the dividing ridge crossing from the Wabash to the Mississippi, with an elevation of about 150 meter above the country both sides.
There are no prairies in this district of tertiary formation, and the common forest trees are such that do not occur north of the dividing ridge,
or only rarely, and not far northward, but all are common on the Ohio

upward, and the same may be said of a number of monocarpic and rhizocarpic species.

Flora Peoriana shall be called that complex of species and varieties of plants that occur around the city of Peoria within a radius of 10 or 12 miles. About 80 per cent. of this area was originally covered by woods, original prairie was little, and there is nearly none now.

THE FOREST.

Generally the forest is mixed of many species, although some are sociable but never excluding others f. i. the white oak and the hickories, with the hazel as undergrowth in the upland, and the willows and cotton wood in the bottoms.

The whole number of woody plants included, the small shrubs is 112 in 58 genera and 30 orders: Ranunculaceæ, 1; Anonaceæ, 1; Menispermaceæ, 1; Tiliaceæ, 1; Anacardiaceæ, 3; Rutaceæ, 2; Vitaceæ, 4; Rhamnaceæ, 3; Celastraceæ, 2; Sapindaceæ, 5; Leguminosæ, 4; Rosaceæ, 15; Saxifragaceæ, 3; Hamameliaceæ, 1; Cornaceæ, 6; Caprifoliaceæ, 6; Rubiaceæ, 1; Ericaceæ, 3; Ebenaceæ, 1; Bignoniacea, 1; Oleaceac, 5; Lauraceæ, 1; Thymelaceæ, 1; Urticaceæ, 4; Platanaceæ, 1; Juglandaceæ, 8; Cupuliferæ, 12; Salicaceæ, 13; Coniferæ, 2; Smilaceæ, 1.

The biggest trees are the white elm, the sycamore, the cottonwood, the soft maple, the white oak and the swamp white oak. They attain often a diameter of 15 decimeter; then follow with an average thickness of 13 decimeter, the scarlet, red and burr oak, the sugar maple, the black walnut, the two shell-bark hickories, the pignut and mockernut, the basswood, the box elder, the honey locust and the hackberry; of 6 to 9 decimeter; the five species of ash, the slippery elm, the butternut, the pecan, the bitternut, the shingle and chestnut oak, the coffeenut, the wild cherry; a thickness of 3 to 6 decimeter attain the mulberry, buckeye, sassafras, the American and large-toothed aspen, black willow and the red cedar.

Small trees are the persimmon, service berry, red bud, paw paw, hornbeam, hop hornbeam, crabapple, sheep berry, plum, buckthorn, hop tree and the haw thorn.

Large shrubs often of tree form with stems 1 decimeter thick are the following: witch hazel, wahoo, prickly ash, bladdernut, long-leaved willow, smooth sumach, the silky, panicled and alternate-leaved cornel, the black haw. Large shrubs are the hazelnut, the false indigo, chokecherry, elder, button bush, rough-leaved and red osier dog wood, arrow wood, aromatic sumach, the silky and glaucous willows.

Very small shrubs are three species of roses: the New Jersey tea, currant and gooseberry; leather wood, the hoary, prairie, dwarf and myrtle willows; black huckleberry, low blueberry, bearberry, wild hydrangea and the alder-leaved buckthorn.

Woody climbers are the trumpet creeper, the Virginia creeper, the poison ivy, the Virgin's bower, 3 species of grapevines, the yellow honey suckle, the waxwork, the climbing rose, the hispid green brier and the Canadian moonseed.

As much mixed as the forest is, the different species of trees and shrubs do not like the same localities. Some prefer a dry, others a moist soil.

The left bank of the Illinois river opposite Peoria is bottom land one mile wide, in spring inundated and for the greater part still thickly wooded. Nearest the bank we find the long-leaved and the black willow, cotton wood, sycamore, soft maple and white elm. Farther backward, black walnut, butternut, pecan, ribbed hickory, hackberry, slippery elm, burr oak, swamp white oak, the five species of ash, of which the green ash is the most frequent, the coffee nut, the honey locust, mulberry, box elder, the buckeye, the paw paw, the persimmon, the elder and the false indigo. On open swampy places, the button bush, the osier dogwood and species of shrubby willows.

The highest trees climb the grapevine, the poison ivy, the Virginian creeper and the trumpet creeper; on shrubs, the climbing rose and the hispid green brier.

Along the foot of the bluff and upward the forest is composed of sugar maple, scarlet, red and shingle oak, chestnut oak, wild cherry, large toothed and American aspen, hop tree, service berry, bass wood, hornbeam, hop-hornbeam, sheeps berry. The undergrowth is composed of witch hazel, wahoo, prickly ash, bladdernut, buckthorn, arrow wood, chokecherry, rough-leaved dog wood, panicled and alternate-leaved cornel, yellow honey suckle and wax work.

On the upland the prominent trees are the white oak, shell-bark hickory, mockernut, bitternut. The brushes hazel, New Jersey tea, smooth sumach and prairie willow.

Coniferous trees are rare: red cedar on single places in small groups and of small growth. The arbor vitae seems to be extinct, but no doubt was once indigenous, for 30 years ago a single tree 60 centimeter thick was found in a swampy place, where it was certainly not planted. The trunk was inclined in an angle of 35 degrees from the ground, so that one could walk up to the branches.

Some of the larger trees grow very rapidly, f. i. cottonwood and sycamore, others have slow growth. A white oak, 11 decimeter thick, was examined, and 250 circles could be counted, 25 of them occupying the sapwood. A sugar maple 92 centimeter thick had 230 circles, and the growth of the last 100 years was only 2 decimeter. The outermost layer of the bark was 125 years old, and the width of the bark was 3 centimeter. A black walnut 75 years old was 3 decimeter thick — a thickness attained by cottonwood already in 20 years.*

---

*The biggest tree I measured was a bald cypress, in Pulaski county. It had in 4 feet from the ground a diameter of 2.1 meter.

Trunks, branchless to a great height, are not rarely found of cotton wood and sycamore, and then the crown has a table form, although the first growth is pyramid-shaped. The coffeenut tree, the ash, the pecan, the sassafras and the wild cherry have a slender trunk. A wide spreading dome-shaped crown have the box elder, the elm, the basswood, the sugar maple and the honey locust, which is conspicuous by its slender horizontal branches. Few branches only have the walnut, butternut, hickories and the coffeenut. Very irregular and knotted-branched is the white oak; densely-branched the elm, hackberries and the shingle oak.

The form of the leaves is manifold: 13 species have lobed leaves: all the oak (except the shingle oak and the chestnut oak, which is only toothed), the maples, sycamore, etc. The leaves of a number are deeply divided or palmate as the buckeye, and not less than 22 species, of which 16 large trees have pinnately divided leaves: the ash, hickories, walnut, butternut, coffeenut and honey locust. The last often very large trees with long branches and double pinnate leaves and small leaflets, represent strength joined with elegance.

The forest is adorned by various tints, changing with the seasons. In March, wherever the elm is predominant, it appears reddish brown, for the white elm (and the soft maple) is the first in bloom. Then follow the male trees of cotton wood with dark-red catkins. In April, the plum and serviceberry with white, the red bud with peach red, and the sugar maple with yellow blossoms. All these develop the flower before leafing. The first young green in April show the buckeyes, and soon afterwards the upright yellow bunches of flowers open in the last days of April or the first week in May. About the 10th of May the forest is green all over; only the sycamore extends the whitish branches leaflets, for that is the last of all the trees that does develop the leaves.

Most of the other trees and shrubs are in bloom during the rest of May. The latest are in middle June the coffeenut, and at the end of June or in the first days of July the basswood, and, finally, the witch hazel opening the flowers only in November, when the leaves are withered and fallen and the fruits of the last year elastically disperse their seeds. Some of the conspicuously blooming shrubs and climbers adorn the woods and copses in July: the elder, the climbing rose, the Virgins bower and the trumpet creeper.

In the fall the forest is shaded of manifold tints by changing the green of the leaves into deep red *(Rhus glabra* and *Ampelopsis)* light red *(Quercus rubra* and *coicinea)* orange, (sugar maple) yellow, *(Prunus serotina* and *Amelanchier)* and brown of every shade *(Platanus* and many others). The hickories are the first that shed the leaves; often already at the end of August. In the latter part of October most of the trees are leafless. Only the white oak and the shingle oak keep partly the dry leaves all winter.

The forest is rich in herbaceous plants (about 250), many of them with conspicuous flowers of vivid colors. The first in flower are Hepatica and Trillium nivale, mostly in the middle of March, and the last in fall are most of the compositæ, species of Aster, Solidago, Helianthus, Eupatorium, Helenium, etc.

THE PRAIRIE.

The largest prairie of the district in question does not exist any more. It was where now the city of Peoria is built, and the little that is left is no more a prairie, for most of the prairie plants are replaced by immigrated foreign weeds. The smaller prairies of the district are turned mostly into cultivated land; still most of the prairie plants that occur in Illinois are represented.

How originated the prairie? To solve this problem many attempts were made. Violent storms were accused to prevent the growth of trees. But the wooded districts are exposed to the same violent storms that prostrate the largest trees but never the young pliant stems, which will grow when the older ones are gone.

There is a general belief that the annual prairie fires make the prairie; a view which without hesitation even Volney adopted in his "Tableau du climat et du sol des Etats unies" ("ou la nature du sol et *plus encore* les incendies anciens et annuel des sauvages ont *occassioné* de vastes deserts"). A burnt forest will not be turned to a prairie, for at once brambles and other shrubs grow up and defend the forest soil against the invasion of the prairie, and we find no remnant stumps below the sod what would prove that there was once a forest.

Prevailing dry winds and deficiency of rain may be the cause of treeless tracts in southern Russia, but certainly not in the upper Mississippi valley with an annual average of 35 inches of precipitation with about one hundred rainy days. To explain the origin and existence of the prairies we have to resort to geological causes.

Lesquereux, in Geological Survey of Illinois (I., 238-254), demonstrates that the soil of the prairie was formed under water by slow decomposition of water plants. When the land emerged it was first swamp, then wet prairie, at last dry prairie with a compact sod of grasses; the soil had such physical and chemical qualities that no tree could grow. To make trees grow it is necessary to plow deep and to expose the soil, so rich of ulmic acid, to the atmosphere. Bushels of tree seeds may be thrown on the surface of the prairie, they will not germinate; or even if they germinate the roots will not penetrate the sod.

Of all the theories this is one of the most acceptable. But there are often small prairies in the midst of forests that have a quite different origin. They are the result of the work of the beaver that build a dam

5

across a brook; ponds were formed, the trees died when such that could not exist in water; afterwards, when the beaver was gone, by and by the dam broke, the water flew off and grasses formed a meadow. Such forest meadows may sooner or later turn again into forest and have nothing in common with the true prairies.

In the upper Mississippi district forest and prairie struggled for existence before culture ended that struggle by subdoing both to the plow. Now the trees are planted on the cultivated prairie land, and so, although regardlessly wasted even in localities fit only for tree growth, the forest got the advantage over the prairie. But even before, when forest and prairie were in their natural state, the trees gained ground at the cost of the prairies. When we examine the constituent parts of the western woods, we find that they gradually diminishing in species follow the large river vallies and even along the small tributaries; then we conclude that all those species from their eastern home traveled westward. The fittest to travel and to settle: the cottonwood, the sycamore and the hackberry, the elm spread the farthest towards the Rocky Mountains, others not farther than Missouri and Iowa, or did not pass over the Mississippi. Not the woods, as was believed, yielded to the prairie, but, on the contrary, the prairie yielded to the woods.

When in the course of time the wide level plain was furrowed by streaming water, from year to year hollowed banks sank down, deep vallies were formed, flanked with steep bluffs once the banks of the rivers, then sod and argillaceous underground and humus thoroughly mixed, as to-day by the plow on the prairie, formed a soil fit to receive the spreading forest growth, and species after species could migrate to the far west.

As the woody plants did westward, so the prairie plants wandered eastward, diminishing by and by in number. Of 55 species of the prairie-flora, that under the same latitude don't go beyond the Alleghanies, 23 do not reach the State of Ohio, and Iowa has many western species that eastward do not cross the Mississippi.

We distinguish wet and dry prairie. The former in the river bottoms or in depressions of the dry prairies which occupy the high and undulating plain.

The number of species of our prairie plants is scarcely more than 200, and many of them are not restricted to the prairie. The first in spring blooming on the dry prairie are: Draba caroliniana, Anemone decapetala, Ranunculus fascicularis, Oxalis violacea, Androsace occidentalis. Then follow in May, Lithospermum angustifolium,* canescens and hirtum, Troximon cuspidatum, Baptisia leucophaea, Pentstemon pubescens; in June, Viola delphinifolia, Scutellaria parvula, Linum sulcatum, Polygala incarnata and sanguinea, Asclepias Meadii and obtusifolia, Sisyrinchium Bermudiana, Tradescantia virginica, Cirsium pumilum, Silene Antirrhrina,

---

*Lithospermum longiflorum Spreng. is the same as L. angustifolium Mich. and was founded on specimens with earlier and larger flowers.

Cacalia tuberosa; in July, Silphium laciniatum, terebinthinaceum and integrifolium, Echinacea angustifolia, Coreopsis palmata and lanceolata, Rudbeckia hirta and subtomentosa, Lepachys pinnata, Asclepias tuberosa and verticillata, Euphorbia corollata, Petalostemon violaceum and candidum, Amorpha canescens, Desmodium Illinænse, Ruellia ciliosa, Callirhœ triangulata, Potentilla arguta and Erynchium yuccæfolium; in August, Helianthus rigidus and occidentalis, Solidago rigida and missouriensis, Hieracium longipilum, Diplopappus linariifolius, Liatris cylindracea, scariosa and pycnostachya, Prenanthes aspera, Gnaphalium polycephalum, Chrysopsis villosa; lastly in September, Aster sericeus, azureus, oblongifolius, multiflorus and ericoides and Gentiana puberula. Number, size and color make these most characteristic members of the prairie flora conspicuous; besides ought to be mentioned a number of tall gregarious grasses: Chrysopogon nutans, Andropogon furcatus and scoparius, Kœleria cristata, Eatonia obtusata, Elymus canadensis Stipa spartea and Sporobolus heterolepis.

The flora of the wet prairies is mixed with the species of the neighboring banks, swamps and bottom-wood and differs according to the soil (sand or silt). The most conspicuous species are in spring: Menyanthes trifoliata, Scilla Fraseri, Allium canadense; in summer, Spiræa lobata, Phlox glaberrima, Asclepias Sullivantii, Sagifraga pennsylvanica, Phaseolus diversifolius, Steironema longifolia and lanceolata, Ipomoea lacunosa, Habenaria leucophaea; in fall, Boltonia asteroides (glastifolia) Prenanthes racemosa, Solidago neglecta, Riddellii and ohioensis, Helianthus giganteus Gentiana Andrewsii and Polygonum ramosissimum — the latter often six feet high. Besides there are many Cyperaceæ: Scirpus lineatus and atrovirens, Cyperus erythrorhizus, Michauxianus and strigosus, Carex vulpinoidea, crus corvi, stipata, conjuncta, arida, scoparia, straminea, cristata, hystricina, tentaculata and some tall grasses: Calamagrostis canadensis, Leersia lenticularis and Spartina cynosuroides.

WATER, SWAMP AND MOIST PLACES.

From the foot of the eastern bluffs numerous springs rush in small beds toward the lower bottom land or form swampy places of little extension. There we find Caltha palustris, Cardamine rhomboidea, Parnassia caroliniana, Archangelica atropurpurea, Aster corymbosus, umbellatus and infirmus, Solidago patula, Cnicus muticus, Archemora rigida, Lysimachia thyrsiflora, Gerardia purpurea, Chelone glabra, Seymeria macrophylla, Symplocarpus fœtidus, Peltandra virginica, Habenaria hyperborea, Cypripedium candidum and spectabile; in the running water: Mimulus Iamesii Ludwigia palustris, Berula angustifolia, Veronica Anagallis and americana, Callitriche heterophylla and Anacharis canadensis.

In the stagnant water of an artesian sulphur well grows abundantly Zannichellia palustris, before the boring of this well never observed in our vicinity.

The species of the larger open swamps are: Epilobium palustre and coloratum, Elodes virginica, Proserpinaca palustris, Cicuta maculata and bulbifera, Sium cicutæfolium, Aster puniceus, Novi Belgii, junceus, salicifolius, paniculatus, Coreopsis aristosa Lobelia Kalmii, Utricularia intermedia, Pedicularis lanceolata, Polygonum sagittatum, Rumex orbiculatus and verticillatus, Acorus calamus, Typha latifolia, Triglochin palustre and maritimum, Dulichium spathaceum, Eleocharis palustris, Eriophorum gracile, Rhynchospora alba, Carex polytrichoides, teretiuscula, filiformis, comosa, riparia, monile, Muhlenbergia glomerata, Phragmitis communis, Phalaris arundinacea, Aspidium Thelypteris and Osmunda regalis. The two latter in a cold bog together with Salix candida and myrtilloides.

In stagnant water of the sloughs where the silt is covered by about a foot deep of water we find: Ranunculus multifidus, Utricularia vulgaris, Saggittaria variabilis Sparganium eurycarpum, Alisma Plantago, Scirpus validus, Potamogeton pauciflorus, Polygonum amphibium, Lemna minor, polyrrhiza and trisulca, Nymphæa tuberosa, Nelumbium luteum, Pontedera cordata and Zizania aquatica. The latter sometimes 4 meter high.

The running water of creeks and shallow places in the river is the abode of Schollera graminea, Valisneria spiralis, Najas flexilis, Ceratophyllum demersum, Nasturtium lacustre, Potamogeton natans, pectinatus and pusillus.

Wet sandy banks harbor with preference: Clematis Pitcheri, Desmanthus brachylobus, Corydalis aurea, Polauisia graveolens, Conobea multifida, Euphorbia heterophylla, Echinodorus rostratus, Cyperus diandrus, inflexus and phymatodes, Hemicarpha subsquarrosa, Fimbrystilis autumnalis, Scirpus pungens, Eragrostis reptans, Frankii and capillaris.

Where the banks are flat and miry, there are: Nasturtium palustre and sessiliflorum, Cardamine hirsuta, Gratiola virginiana, Bidens chrysanthemoides and connata, Eclipta alba, Ilysanthes gratioloides, Lippia lanceolata, Sagittaria heterophylla, Iris versicolor, Eleocharis obtusa and acicularis.

Along brooks we find: Thalictrum cornuti, Silene nivea, Hypericum pyramidatum, Hibiscus militaris, Thaspium barbinode, Artemisia Ludoviciana, Erigeron philadelphicum, Solidago lancelota and tenuifolia, Silphium perfoliatum, Cacalia suaveolens, Plantago cordata, Scrophularia nodosa, Carex shortiana, Leersia oryzoides, Glyzeria nervata and fluitans, Panicum virgatum.

ROADSIDES, WASTE PLACES, PASTURES AND CULTIVATED SOIL.

Of roadsides, wastes places in the neighborhood of houses and yards, mostly immigrated plants took possession, but some of the indigenous ones kept their places: Lepidium virginicum, Mollugo verticillata, Erigeron canadense, Ambrosia artemisiaefolia, Dysodia chrysanthemoides, Bidens frondosa, Erechtides hieracifolia, Artemisia biennis, Verbena stricta, urticifolia and bracteosa, Solanum carolinense, Datura Tatula, Polygonum aviculare and erectum, hydropiper and pennsylvanicum, Hordeum pratense and Eragrostis pectinacea.

On pastures prevail Trifolium repens, Poa pratensis and compressa; and on sandy places Cyperus filiculmis, Vilfa aspera and vaginaeflora, Panicum autumnale.

Along fences many tall plants are preserved: Napaea dioica, Gaura biennis, Oenothera biennis, Ambrosia trifida, Helianthus grosseserratus and doronicoides, Lactuca canadensis and Asclepias cornuti.

The indigenous weeds between the cultivated plants are: Sisymbrium canescens, Potentilla norvegica, Erigeron annuum, Xanthium canadense, Veronica peregrina, Ipomoea pandurata, Physalis virginiana and lancolata, Chenopodium album and hybridum, Euphorbia maculata and hyperici folia, Panicum capillare and crus galli, Cenchrus tribuloides.

IMMIGRATED PLANTS.

Before this country was settled by our race, at the time when the red man hunted the buffalo on the prairie and the elk in the forest, the virgin soil was intact, nature reigned undisturbed. When the white man came with his plow, he introduced voluntary or involuntary many foreign plants, which by and by spread and supplanted indigenous plants on the cultivated land.

It would promote phytogeography, when for each local flora that immigration could be historically verified and transmitted to posterity. In regard to our flora an attempt shall herewith be made as far as possible.

The species which immigrated partly from Europe, partly from tropical countries, are either fully naturalized and form an integral part of our present flora, or they are adventives, *i. e.*, new comers, mostly escaped from cultivated land or gardens, that may afterwards become naturalized, after a more or less prolific propagation, or become extinct again, when the chances are less favorable.

Perfectly naturalized and common around Peoria in the year 1852 were: Sisymbrium officinale Scop., Brassica nigra L., Capsella pursa pastoris Moench, Portulaca oleracea L., Malva rotundifolia L., Sida spinosa L.,

Abutilon Avicennæ Gærtn., Trifolium pratense L., Pastinaca sativa L., Maruta cotula DC., Lappa officinalis All., Verbascum Thapsus L., Chenopodium urbicum L., Chenopodium botrys L., Chenopodium ambrosioides L., Polygonum persicaria L., Polygonum Convolvulus L., Rumex crispus L., Cannabis sativa L., Phleum pratense L., Eragrostis poæoides Beauv. var megastachya, Eragrostis pilosa Beauv., Bromus secalinus L., Panicum sanguinale L., Setaria glauca Beauv.

Old settlers, but not so common, are: Hypericum perforatum L., Veronica arvensis L., Nepeta cataria L., Nepeta glechoma Benth., Marrubium vulgare L., Melilotus alba Lam., Malva sylvestris L., Martynia proboscidea Glox., Amarantus spinosus L., Rumex obtusifolius L., Rumex acetosella L., Dactylis glomerata L., Panicum glabrum Gaud.

First observed between 1855 and 1860, and now very common: Sonchus asper Vill., Linaria vulgaris Mill.. Leonurus cardiaca L., Echinospermum Lappula Lehm., Cynoglossum officinale L.

New settlers atter 1860 and before 1870 and now common: Nasturtium officinale R. Br., Stellaria media, Smith; less common: Verbascum Blattaria L., Melilotus officinalis Willd., Eleusine indica Gärtn., Setaria verticillata Beauv. Trifolium arvense L. Lychnis Githago Lam., Camelina sativa L. Cirsium arvense Scop.

Single specimens were collected 1852 (but not seen since) of: Raphanus Raphanistrum L., Nicandra physaloides Gærtn., Inula Helenium L and Leucanthemum vulgare L.

The latter re-appeared 1885 on the railroad tracks.

In 1886 first appeared Conium maculatum L., and Lactuca scariola L.

Several years ago single stocks of two grasses appeared along railroad tracks, but not seen since: Triticum repens L and Lolium perenne L. Probably the seed dropped from the cars but did not propagate.

Some species sometimes escape from gardens or cultivated lands, but are, until now, not naturalized: Argemone mexicana L., Nasturtium Armoracia Fr., Hibiscus Trionum L., Medicago sativa, Rosa rubiginosa L., Anethum graveolens L, Daucus carota L, Helianthus annuus L., Tanacetum vulgare L., Centaurea Cyanus L., Mentha viridis L., Satureja hortensis L., Ipomoce purpurea Lam., Ipomoea Nil Roth, Lycium vulgare Dun., Polygonum orientale L., Fagopyrum esculentum Mœnch, Asparagus officinalis L., Phalaris canariensis L and Setaria Italica Kunth.

There is a number of species which are doubtful whether indigenous or naturalized,

Cerastium triviale Link (C. vulgatum L. spec., C. viscosum L. herb et auct. americ.) is the first time mentioned by Pursh (Flora Am. sept. I., 320) and then by Nuttall (Gen. of N. Am. pl. I., 291), neither of them says whether the plant be introduced or not. Barton (Fl. phil. I., 216) says that perhaps only C. longepedunculatum Muhl (C. nutans Raf.) and C. arvense L are indigenous. Decidedly as an introduced plant it is spoken of by Beck (Bot. of N. St. 51) by Dewey (Herbaceous plants of Mass. 89) by Darlington (Fl. cestr. 33) and Torrey (Flora of N. Y. I., 99). In Torrey and Gray's Flora I., 188 it is left doubtful whether introduced or not, and Gray in Man. of Bot. says: perhaps indigenous to the country. Not a few plants, common to North America and Europe, are supposed to have come to the eastern states on a double way by new introduction from Europe or by old migration from the north in both continents. The question is, whether this in regard to our plant is possible and probable.

A. de Candolle in Geogr. bot. 748 says: Cerastium vulgatum L. et Cerastium viscosum L paraissent manquer encore à l'Asie orientale et au nord-ouest de l'Amerique, ce qui me fait croire à l'introduction aux États-unies.

The plant is found in Iceland (Hooker Tour in Iceland II., 324) in Labrador (Meyer Flor. Labr. 94) in Spitzbergen (Martens flor. arct. 8) in Greenland (Lange in Etzels Greenland 634). So an arctic connection with the eastern continent is proved. After Ledebour (Fl. ross. I., 408), who takes the species in a wide sense, it extends through Sibiria to Kamtschatka, and his varieties Grandiflorum and Behringianum to Alaska. The latter after Watson (Kings Rep. V., 38) occurs in the Uintah Mountains in an altitude of 10,000 feet, and in the Rocky Mountains it was found by Parry (Pl. of Rock. Mount. in Proc. Ac. n. s. Phil. 1863, p. 55) and Gray in Pl. Wright, I., 18, mentions: Cerastium vulgatum L., C. triviale Auct,, in ravines of the Organ Mountains N. Mex. That means, no doubt, L. spec. and not L. herb as in Man. of Bot.

Now, if a plant is distributed from the arctic region to New Mexico, it is possible that it spread in the same direction to Illinois. It is not mentioned in Agassiz Lake Superior, but Houghton collected there C. viscosum L. (Schoolcraft Sources of the Mississippi), and we find it in Lapham's Catalogue of Wisconsin plants. In both cases it is not said whether L. herb or spec. is understood.

I found the plant on a grassy place at the bank of a little brook at that time remote from a settlement, what made me suppose, that the plant did not recently immigrate.

Solanum nigrum L. is a polymorphus cosmopolite, and now ranged amongst the indigenous by Gray in Fl. of N. Am. I., 227.

Datura Tatula L.—The question, whether this is a proper species or a variety, and the probability of an American origin, A. de Condolle in Geogr. Bot., 731, has extensively treated. Another question is, whether we have the plant from South America, as is supposed. When that is so, the immigration must be a very old one. Western farmers affirm that in remote new settlements the plant appear as soon as the land is broken. It is a well known fact, that the seeds of most Solanaceæ keep the germinative power a long time, and so it is probable that the seeds had been buried in the ground long before they had a chance to germinate. The white species Datura stramonium was, only a few years ago, the first time seen — in the yard of a druggist's house !

Chenopodium album L. and hybridum L.—The American botanists took those plants unanimously for immigrated ones. Amongst the fifty-five species described by Mocquin in Prodromus XIII., 2, are only two, the exclusive habitat of which is conceded to North America, one in California, the other in Carolina and Texas; the species of the genus are equally distributed all over the globe, and the species in question are cosmopolites, the original home unknown; they may have spread by colonization in settled countries, but recently both species were found in the Rocky Mountains of Colorado in an altitude of 10,000 feet (Porter & Coulter, Flora of Colorado, 110), C. album L. from the Great Bear Lake, $66°$ N. L. to Nevada, and C. hybridum L. from the Saskatchewan to the Wahsatch Mountains. Watson (in King's Report, V. 287), declares both to be indigenous, and in his Revision of N. Am. Chenopodiaceæ (Pr. Am. Ac. IX, 97), under C. hybridum, he says: "Introduced eastward, but indigenous from Kentucky, Texas, and New Mexico to Oregon." I never doubted that the variety viridis be indigenous, as it is found in our vicinity only in the shade of the woods.

Amarantus retroflexus L. and albus L.—These species, too, were ever taken for immigrated plants. To Mocquin it is an open question, whether the former came to Europe from America, as to the latter he is sure of it: he gives as habitat Pennsylvania and Virginia. In the meantime the plants were found in the deserts beyond the Rocky Mountains, "far from cultivated fields," as Watson remarks in King's Report, and "probably indigenous." De Condolle did not include them in the list of immigrated plants in Geogr. Botan.

There is a third species, A. blitoides Wats., which I first observed about ten years ago, when I took it for a variety of A. albus. Now it grows in immense numbers around houses. Watson described it as a new species (Flora of California, II., 41), "Mexico to Northern Nevada and Iowa, and spreading then eastward." As its migration seems to be spontaneous and steady, and being a North American plant, I ranged it in the list of indigenous plants.

Poa annua L. Poa compressa L. and Poa pratensis are no doubt circumpolar. They occur from Europe through Siberia to Kamtschatka, Poa annua in Sitka (Alaska), after Ledebour (Flora Rossica IV., 372–78), in Greenland and Iceland, after Martens (Fl. arct.), P. Pratensis in Greenland, Iceland and Labrador, Poa compressa in Labrador, after Meyer (Plant· Labrad., 19). So we may have the plants from the North as well as from the East.

Panicum crus Galli L. is a cosmopolite, the original home of which to verify will hardly be possible; it occurs throughout North America, even in the deserts of Utah and Nevada.

Agrostis alba L. and A. vulgaris With., have the same distribution as the three species of Poa. They are in Gray's Manual acknowledged as indigenous.

CULTIVATED PLANTS.

The foreign, not American, trees and shrubs, planted for shade or ornament around houses, are: Aesculus Hippocastanum, Ailanthus glandulosus, Populus alba and dilatata, Salix alba and babylonica, Tilia europæa, Philadelphus coronarius, Syringa vulgaris, Ligustrum vulgare, Rhamnus catharticus. Lycium vulgare, Rhus cotinus and Catalpa Kæmpferi.

American species from southern Illinois: Liriodendron tulipifera, Catalpa speciosa, Robinia pseudacacia, Wistaria frutescens; from north Illinois: Betula papyracea; from other States: Populus balsamifera, Pyrus americana, Catalpa bignonioides, Cratægus cordata, Ribes aureum, Symphoricarpus racemosus, Lonicera sempervirens and Maclura aurantiaca.

Coniferæ: Pinus Strobus, Abies balsamea and Larix americana, and the foreign Pinus austriaca and Abies excelsa thrive only in clayey soil of the bluffs.

In tne orchard the apple in many and often excellent varieties takes the first place; less common are cherries, pears and plums. Peaches suffer from severe winters and late frosts. Small fruit for the market are raised: Strawberries, raspberries, gooseberries and currants.

The principal crops of cultivated farm land in Illinois are: Maize, wheat, oats, rye, barley, buckwheat and potatoes, together (in 1875) on 13¼ millions of acres, about 38 p. c. of the area of the State. Hay was made in the same year on 2¼ millions of acres on uncultivated farm land.

Maize alone was cultivated on more than eight millions of acres. Indeed no Illinois farm is so prosperous as the large farm Esperanza, in Mexico, of which Humboldt, in his Essay Politique sur la Nouvelle Espagne, says: "Une faneque de mais en produit quelquefois huit cent." Still the Illinois farmers reap nearly the average crop of the fertile parts of that country, which is, after Humboldt, 3–400 fold, that would be 288 fold after the following table, which I owe to the kindness of Mr. Roswell Bills, Sec. of Peoria Co. Agricult. So.:

	TIME OF SOWING.	TIME OF HARVEST.	QUANTITY OF SEED.	CROP.
Maize	May 14th.	Oct. 10th.	5 quarts,	45 bushels.
Winter Wheat ...	Sept. 1st.	July 1st.	1½ bushels.	20 "
Spring Wheat....	Mch. 25th.	July 15th.	1½ "	15 "
Rye	Sept. 10th.	June 25th.	1½ "	20 "
Oats	April 10th.	July 15th.	2⅓ "	50 "
Barley	?	June 20th.	2½ "	50 "
Buckwheat	June 25th.	Sept. 25th.	½ "	20 "
Potatoes	May 15th.	Oct. 1st.	5 "	100 "

In the year 1875, wheat was grown on 2.6 millions of acres, oats on 2.27 millions of acres, and potatoes on 118,715 acres, with a crop of 128 bushels per acre. The cultivated fodder are clover and timothy.

6

Other farm products are: Peas, beans, sweet potatoes, pumpkins, cucumbers, melons, water melons and tomatoes. What the gardening bring to market is the same as in all civilized countries of the temperate zone.

SYSTEMATIC SYNOPSIS OF THE FLORA OF PEORIA.

To indicate the relative number of localities and the relative number of individuals in each locality, roman figures are used for the former, and arabic ones for the latter; so Eleocharis palustris X., 10, means that the plant occurs on all favorable places, in swampy places, flat banks a. s. o. in the greatest number; Corylus americana VIII, 10, that it is found in most favorable localities; Thuja occidentalis I., 1, that this tree was observed on a single place a single specimen. The abbreviated geographical names* express the limits of distribution of each species.

*ABREVIATED GEOGRAPHICAL NAMES.

[Those of the States are well known, and therefore omitted.]

All—eghannies, All. (N.Ca.), All. (Ga.), Alleghannies southward to N. Ca, or Ga.	Miss—issippi river.
Art—ic coast,	U. Miss.—Upper Mississippi.
Atl—antic coast.	U. Mo.—Upper Missouri.
Can—ada.	N. E.—New England.
G—ulf coast.	N. F.—New Foundland.
Hud—son.	Pac—ific coast.
Labr—ador.	R. Mts.—Rocky Mountains.
	Sask—atchewan.

ABBREVIATED NAMES OF AUTHORS.

Ach—arius.	Dun—al.	Lindl—ey.	Schreb—er.
Ait—on.	Ehrh—art.	Lk.—Link.	Schrad—er.
Anders—son.	Ell—iot.	Lightf—oot.	Schk—uhr.
Andr—ews.	Englm—ann.	L'Her—itier.	Schult—es.
Aust—in.	F.& M.Fischer& Mayer	Lois—eleur.	W. P. Sch—imper.
Bart—on.	Fr—ies.	Marsh—all.	Schpr—Schimper.
Beauv—ois.	Frœl—ich.	Mart—ins.	Schwæg—richen.
Benth—am.	Gærtn—er.	Mey—A. C. Meyer.	Schw—einitz.
Bernh—ardi.	Gaud—ichaud.	Michx—Michaux.	Scop - oll.
Big—elow.	Gm—elin.	Mill—er.	Sh. & P.—Short and
Bisch—off.	Good—enough.	Mocq—uin.	Peter.
Bland—ow.	A. Gr—ay.	Muell—er.	Shuttl—eworth.
A. Br—aun.	Grev—ille.	Muhel—enberg.	Sibth—orp.
R. Br—own.	Gris—ebach.	Murr—ay.	Sm—ith.
Brid—el.	Gron—ovius.	Norm—ann·	Spr—engel.
Br. Sch. Bruch and Schimper.	Hassk—arl.	Nutt—all.	Spring—er.
Br. Eu. Bryologia Europea.	Hedw—ig.	Pers—oon.	Sol—ander.
Buckl—ey.	Hoffm—ann.	Ph.—Pursh.	Steud—el.
Cav—aniles.	Hook—er.	Planch—on.	Sull—ivant.
Chois—y.	Hornsch—uch.	Poir—et.	Sw—artz.
Curt—is.	Hueb—ener.	Raf—inesque.	Trin—ius.
Darl—ington.	Hud—son.	Rich—ard.	Torr—ey.
DC.—DeCandolle.	H. B. K. Humboldt Bonpland & Kunth.	Richards—on.	T. Gr. — Torrey and Gray.
ADC.—Alphonse DC.	Jacqu—in.	Ridd—ell.	Tuck—erman.
Desn.—Decaisne.	Juss—ieu.	Rostk—ovius.	Vent—enat.
Desf—ontaine.	Kth.—Kunth.	R. Sch. — Rœmer and Schultes.	Wang—enheim.
Dew—ey.	Lag—asca.	R. P.—Rulz & Pavon.	Wahl—enberg.
Dietr—ich.	Lam—ark.	Rottb—oll.	Walt—er.
Dgl.—Douglas.	Lamb—ert.	Salisb—ury.	Wats—on.
Dum—ortier.	Lehm—ann.	Sartw—ell.	Willd—enow.
	L—innæus.	Schleich—er.	With—ering.

Ranunculaceæ. In N. Am. 18 Gen. 147 Spec.

Clematis Pitcheri T. Gr. sandy banks V. 5 Ill.—N. Mex.

Clematis virginiana L. shady banks III. 2 Atl.—N. Mex. G.—Sask.

Anemone decapetala L. dry prairies, hill-sides III. 4 Ariz.—N. Ca.—Utah—Ill.

Anemone cylindrica Gr. copses V. 4 Mass.—Ill. Wisc.—N. Mex.

Anemone virginiana L. woods V. 5 N.E.—Ark. All. (S.Ca.)—55°N.L.

Anemone dichotoma L. bottom wood V. 7 Can.—All. (Pa.) Ark. R. Mts.—Arct.

Hepatica triloba Chaix var. acutiloba wood VI. 5 Atl.—Alaska G.—U. Mo.

Thalictrum anemonoides Michx. wooded hill-sides II. 5 Can.—N. Ca. La.—U. Miss.

Thalictrum dioicum L. woods II. 2 N.E.—La.—67°N. L.

Thalictrum purpurascens L. woods, copses III. 3 N.E.—La.—R. Mts.

Thalictrum cornuti L. woods, shady banks IV. 3 Atl.—N. Mex. G.—56°N.L.

Ranunculus multifidus Ph. stagnant water I. 4 N.E.—N. Ca.—La.—R. Mts. Arct.

Ranunculus abortivus L. woods VIII. N.F.—La.—R. Mts. S.Ca.—57°N.L.

Ranunculus recurvatus Poir. moist woods III. 3 Atl.—Or. G.—Labr.

Ranunculus fascicularis Muhl. prairies and wood openings VI. 6 N. E.—R. Mts.

Ranunculus repens L. woods X. 6 All.—Pac. G.—Can.

Isopyrum biternatum T.Gr. shade VI. 5 G.—Oh.—U.Miss.

Caltha palustris L. springy places in bottom woods III.6 Atl.—Pac. S.Ca.—Can.

Aquilegia canadensis L. wooded hillsides VI. 3 Atl.—N.Mex. Ga.—Huds.

Delphinium tricorne Michx. copses IV. 2 All.—Up.Mo.—La.

Hydrastis canadensis L. shady woods II. 3 Can.—Ga.—Up.Miss.

Actaea alba Big. shady woods V. 3 N.E.—La.—55° N.L.

Anonaceæ. In N. Am. 2 Gen. 5 Spec.

Asimina triloba Dun. bottom woods V. 4 All.—Miss. G.—Gr. Lakes.

Menispermaceæ. In N. Am. 3 Gen. 5 Spec.

Menispermum canadense L. bottom woods Atl.—R.Mts. G.—Can.

Berberidaceæ. In N. Am. 7 Gen. 15 Spec.

Caulophyllum thalictroides Michx. shady woods IV. 3 N.E.—N.Ca.—Up. Miss.

Jeffersonia diphylla Pers. shady woods I. 4 All. (Tenn.)—Mich.—Ill.

Podophyllum peltatum L. woods VIII. 6 Atl.—Up.Mo. G.—Can.

Nymphæaceæ. In N. Am. 5 Gen. 13 Spec.

Nelumbium luteum Willd. waters III 10 All.—Miss. G.—Can.

Nymphæa tuberosa Paine waters III. 6 Atl.—Miss.

Nuphar advena Ait. waters I. 5 Atl.—Pac. G.—Can.

Papaveraceæ. In N. Am. 12 Gen. 17 Spec.

Sanguinaria canadensis L. woods VI.5 Atl.—Miss G.—Can.

Fumariaceæ. In N. Am. 3 Gen. 17 Spec.

Dicentra cucullaria D.C. wooded hillsides VI. 5 N.E.—N.Ca.—Up. Miss.

Corydalis aurea Willd. sandy banks, fields II. 6 Atl.—W.Tx. R.Mts. G—64° N.L.

Cruciferae. In N. Am. 37 Gen. 217 Spec.

Nasturtium sessiliflorum Nutt. banks bottom III. 4 Fla.—Tex.—Ill.

* †Nasturtium sinuatum Nutt. bottom I. 1 Miss.—Pacif.

Nasturtium palustre D.C. bottom VIII. 6 N.E.—La.Pacif.—Arct.

Nasturtium lacustre Gr. inundated bottom II. 4 Can. N.Y. Ky. La.—Up. Miss.

Dentaria laciniata Muhl. woods V. 5 Fla.—La. N.E.—Up.Miss.

Cardamine rhomboidea D.C. bottom, springy places V. 7 Atl.—R.Mts. G.— Can.

Cardamine hirsuta L. bottom VIII. 8 Atl.—Pacif. G.—Arct.

Arabis dentata T.Gr. bottom III. 3 N.Y.—Tenn.—Up.Miss.

Arabis canadensis L. shady hillsides N. 4 N.E.—Ark.—Can.

Arabis laevigata D.C. rocky hillsides N. 4. Can.—Va. N.E.—Up.Miss.—Ark.

Thelypodium pinnatifidum Wats. bottom V. 5 Oh.—Ill.—La.

Sisymbrium canescens Nutt. fields V. 8 N.Y.—Fla.—La.—Cal.—Arct.

Draba caroliniana Walt. hillsides V. 8 N.E.—Ga.—La.—Up.Miss.

Lepidium virginicum L. roadsides, waste places X. 8 Atl.—La.—Up.Mo.

Capparidaceæ. In N. Am. 9 Gen. 25 Spec.

Polanisia graveolens Raf. sandy banks IV. 6 Can.—Fla.—Ark.—Up.Miss.

Violaceæ. In N. Am. 2 Gen. 33 Spec.

Jonidium concolor B.II. shady woods II. 3 N.Y.—All.—Ill.

Viola cucullata Ait. woods and prairies VIII. 5 Atl.—Pacif. G.—Arct.

Viola cucullata var.palmata Gr. woods I. 4.

Viola sagittata Ait. dry openings III. 4 Atl.—Miss. G.—Can.

Viola delphinifolia Nutt prairie III. 5 Ill.—Up.Mo.

Viola pedata L. dry openings 11. 5 Atl.—Miss. G.—53° N.L.

—— var. bicolor.

Viola pubescens Ait. woods VI. 4 N.E.—All. (Ga.)—R.Mts.

Cistaceæ. In N. Am. 3 Gen. 17 Spec.

Helianthemum canadense Michx. open woods, hillsides II. 4 Atl.—Miss. G. —Can.

Lechea major Michx. dry woods, hillsides III. 5 Atl.—Miss. G.—Can.

Lechea minor Lam. dry woods, hillsides II. 8 Atl.—Up.Mo. G.—Can.

Hypericaceæ. In N. Am. 3 Gen. 35 Spec.

Hypericum pyramidatum Ait. banks II. 3 N.E.—Ill. Wisc.

Hypericum corymbosum Muhl. copses V. 5 N.E.—N.Ca.—Miss.

Hypericum sphærocarpum Michx. banks II. 4 Oh.—Ill.—La.

Hypericum nudiflorum Michx. copses V. 5 N.E.—Up.Miss. Fla.—La.

Elodes virginica Nutt. swampy bottom I. 2 N.E.—Up.Miss. Fla.—La.

Caryophyllaceæ. In N. Am. 12 Gen. 129 Spec.

Silene stellata Ait. copses VI. 4 N.E.—All.—La.—Up.Miss.

Silene nivea D.C. bottom II. 4 Pa.—Ill.

Silene antirrhina L. dry hills V. 3 Atl.—Cal. Or. Tex.—Can.

Stellaria longifolia Muhl. grassy places II. 6 N.E.—Va.—La.—R.Mts.—Alaska.

Arenaria lateriflora L. shady hillsides I. 4 N.E.—Up.Mo. Or. Alaska.

Cerastium triviale Link. (viscosum L. herb) grassy places I. 3 N.Engl.—Fla. —N.Mex.—Alaska.

Cerastium nutans Raf. moist places VII. 4 Vt.—Minnes. N.Ca.—N.Mex.— Cal.

Paronychieae. In N. Am. 5 Gen. 17 Spec.

Anychia dichotoma Michx. openings VI. 8 N.E.—Up.Miss. Ga—Ark.

Portulacaceae. In N. Am. 9 Gen. 44 Spec.
Claytonia virginica L. openings VI. 6 Atl.—N.Mex. Ga—Can. Ariz.—
Alaska.

Malvaceae. In N. Am. 17 Gen. 104 Spec.
Callirhoe triangulata Gr. prairies II. 4 All. (Ala—N.Ca)—Up.Miss.
Napæa dioica L. fences, open places, copses III. 5 All.—Ill.
Hibiscus militaris Cav. bottom II. 4 All. (Pa.—Ala.)—Miss.

Tiliaceae. In N. Am. 2 Gen. 4 Spec.
Tilia americana L. woods V. 3 N.E.—All—La.—52°N.L. ·

Linaceae. In N. Am. 1 Gen. 17 Spec.
Linum sulcatum Ridd. prairies, open woods, V. 5 N.E.—N.Ca.—Up.Mo.

Geraniaceae. In N. Am. 6 Gen. 27 Spec.
Geranium maculatum L. open woods V. 5 Atl.—Up.Mo. G.—Can.
†Floerkea proserpinacoides Willd. bottom I. 4 N.E.—Utah.
Impatiens pallida Nutt. moist, shady woods II. 5 Atl.—Oreg. Ga—Can.
Impatiens fulva Nutt. moist shady woods VI. 5 Atl.—Oreg. G.—66°N.L.
Oxalis violacea L. dry hillsides VI. 3 Atl.—N.Mex. G.—Can.
Oxalis stricta L. bottom, cultivated land VIII. 5 Atl.—Pacif. G.—Can.

Rutaceae. In N. Am. 5 Gen. 11 Spec.
Zanthoxylum americanum Mill. copses VIII. 4 N.E.—Va.—Up.Miss.
Ptelea trifoliata L. woods, banks VI. 3 Lake Erie—N.Mex. Fla.—Tex.

Anacardiaceae. In N. Am. 2 Gen. 14 Spec.
Rhus glabra L. dry hillsides VIII. 5 Atl.—Oreg. G.—Sask.
Rhus toxicodendron L. bottom woods VIII. 5 Atl.—Oreg. G. -Sask.
Rhus aromatica Ait. hillsides IV. 2 Vt.—Kas. G.—Sask.

Vitaceae. In N. Am. 2 Gen. 15 Spec.
Vitis aestivalis Michx. woods VII. 5 Atl.—Sonora G.—Can.
Vitis cordifolia Michx. woods VII. 5 Atl.—Miss. G. -Can.
Vitis riparia Michx. woods IV. 3 N.E.—Va.—N.Mex.
Ampelopsis quinquefolia Michx. woods VII. 6 Atl.—N.Mex. G.—Can.

Rhamnaceae. In N. Am. 12 Gen. 47 Spec.
Rhamnus lanceolatus Ph. woods V. 3 W.Pa.—Tenn.—Up.Miss.
Rhamnus alnifolius L'Her. swampy places II. 4 N.Y.—Ill.—Huds.—Cal.
Ceanothus americanus L. copses VI. 5 Atl.—R.Mts.—Tex. G.—Can.

Celastraceae. In N. Am. 7 Gen. 15 Spec.
Celastrus scandens L. woods V. 3 All. (N.Y.—N.Ca.) -Up.Miss.—47°N.L.
Euonymus atropurpureus Jacq. woods, copses V, 3 Can. 47 N.L.—Fla.—Mo.

Sapindaceae. In N. Am. 13 Gen. 31 Spec.
Staphylea trifolia L. copses V. 4 N.Ca.—Tenn.—47°N.L.—Up.Mo.
Aesculus glabra Willd. bottom woods V. 4 W.Pa.— Up.Miss.
Acer saccharinum Wang. wood hillsides V. 8 Can. N.E.—All.—La.—Lake
Winnipeg.
Acer dasycarpum Ehrh. bottom woods VII. 9 Can. N.E.—Ga.—Ark.—Wisc.
Negundo aceroides Moench. bottom woods IV. 4 Pa.—Fla.—Cal.—Sask 54°
N.L.

Polygalaceae. In N. Am. 3 Gen. 40 Spec.
Polygala incarnata L. dry prairie II. 4 Pa.—Fla.—Ark.—Wisc.
Polygala sanguinea L. dry prairie, open woods III. 4 Atl.—N.Mex. G.—Can.

Polygala verticillata L. sandy places IV. 6 Atl.—R.Mts. G.—Can.
Polygala senega L. dry wooded hillsides IV. 3 N.E.—All.—(N.Ca.)—Miss.—
     54°N.L.

Leguminosae. In N. Am 67 Gen. 736 Spec.
†Crotalaria sagittalis L. sandy banks I. 2 N.E.—Up.Miss.—Fla.—Sonora.
Trifolium reflexum L. bottom woods II. 3 All.—Ill.—Fla.—Tex.
Trifolium repens L waste places, pastures, roadsides, X. 10 all North
     America.
Psoralea onobrychis Nutt. copses V. 6 All. (S.Ca.) Oh.—Up.Miss.
Psoralea floribunda Nutt. dry prairies, hillsides V. 6 Ill.—R.Mts.—Ark.—
     Sonora.
Petalostemon violaceum Michx. prairies, hillsides V. 4 Mich.—N.Mex.—La.
     —Sask.
Petalostemon candidum Michx. prairies, hillsides, V. 3 Mich.—Sonora.—La.
     —Sask.
†Tephrosia virginiana Pers. dry prairie I. 1 Atl—Miss. G.—Can.
Amorpha fruticosa L. bottom, banks V. 5 Pa.—Up.Mo.—W.Tex.—Fla.—L.
     Winnipeg.
Amorpha canescens Nutt. dry prairie, hillsides V. 6 Mich.—R.Mts.—Ga.—
     Tex.
Astragalus canadensis L. copses V. 5 All. (N.Y.—Ga.)—La.—Or.—58°N.L.
Desmodium nudiflorum D.C. woods I. 3 Atl.—Up.Mo. G.—Can.
Desmodium acuminatum D.C. woods V. 5 Atl.—Up.Mo. G.—Can.
†Desmodium pauciflorum D.C. woods I. 2 Pa.—Ill.—Fla.—La.
Desmodium cuspidatum T.Gr. copses III. 3 Atl.—Up.Mo. G.—Can.
Desmodium canescens D.C. copses VII. 7 Atl.—Miss. G.—Can.
Desmodium Illinoense Gr. prairies, copses V. 3 Ill.
Desmodium Dillenii Darl. copses III. 3 Atl.—Miss. G.—Can.
Desmodium paniculatum D.C. copses, wood openings V. 8 N.E.—Up.Mo.—
     Fla.—W.Tex.
Desmodium canadense D.C. woods, copses IV. 3 Atl.—R. Mts—N.Ca.—54°
     N.L.
Desmodium sessilifolium T.Gr. copses I. 5 Pa. Ky.—Tex.—Up.Mo.—Mich.
Lespedeza violacea Pers. copses V. 6 Atl.—Up.Mo.—Tex. G.—Can.
Lespedeza reticulata Pers. copses III. 4 Atl.—Miss. G.—Can.
Lespedeza capitata Michx. copses, wood openings V. 5 Atl.—Up.Mo. G.—
     Can.
†Vicia americana Muhl. bottom, moist copses I. 2 N.Y.—Ky.—La.—Gr. Bear
     Lake—N.Mex.—Cal.
Lathyrus palustris L. moist copses III. 4 N.E.—N.Ca.—Pacif.—55°N.L. Lab-
     rador.
Phaseolus diversifolius Pers. bottom V. 3 Atl.—Up.Mo.—W.Tex. G.—Can.
†Phaseolus helvolus L. sandy soil I. 2 N.Y.—Ill.—Fla. N.Mex.
Apios tuberosa Moench. bottom copses III. 4 Oh.—R.Mts. G.—Can.
Amphicarpaea monoica Nutt. woods V. 5 Atl.—Up.Mo. G.—Can.
Baptisia leucantha Nutt. copses V. 4 Oh.—Wisc.—Up.Mo. Fla.—Tex.
Baptisia leucophaea Nutt. prairies V. 3 Ga.—Tex.—Wisc.—Mich.
Cercis canadensis L. woods VI. 3 Pa.—Ill.—Cal. Fla.—La.
Cassia marilandica L. bottom VI. 4 Atl.—Mo. G.—Can.
Cassia chamaecrista L. bottom VI. 7 Atl.—N.Mex. G.—Can.
Gymnocladus canadensis Lam. bottom woods V. 3 W.N.Y.—Tenn.—Up.
     Mo.—46°N.L.
Gleditschia triacanthos L. 5 bottom woods V. 5 W.Pa.—Up.Mo. Fla.—La.

Desmanthus brachylobus Benth. sandy banks III. 4 Ill.—Up.Mo. Ky. La.—
W.Tex.

Rosaceae. In N. Am. 35 Gen. 201 Spec.

Prunus americana Marsh. woods, copses V. 4 Atl.—Up.Mo. Tex.—Sask.
Prunus Virginiana L. copses V. 3 Atl.—R.Mts.—N.Mex. G.—67°N.L.
Prunus serotina Ehrh. woods V. 3 Atl.—Up.Miss.—W.Tex. G.—62°N.L.
Spiraea lobata Murr. bottom II. 4 All. (Pa.—Ga. Ill. Mich.
Spiraea Aruncus L. wooded hillsides IV. 4 All. (N.Y.—Ga.)—R. Mts. Cal.
    Sitka.
Agrimonia Eupatoria L. woods V. 3 Atl.—Pacif. G.—Can.
Agrimonia parviflora Ait. woods II. 3 All. (Pa.—N.Ca.)—Ill.—La.
Geum album Gm. woods, copses V. 4 N.E.—Ga.—Ill.
Geum virginianum L. 4 wet prairies I. 4 N. E.—R.Mts.—W.Tex.
Potentilla norvegica L. fields VI. 5 N.E.—N.Mex.—Or.—Alaska—Arct.
Potentilla canadensis L. woods, copses VI. 3 N.E.—All.—La. Can.—Up.Mo.
Potentilla arguta Ph. prairies, copses III. 4 N.E.—R.Mts.—65°N.L.
Fragaria virginiana Ehrh. wooded hillsides VII. 5 Atl.—Cal.—Or. G.—65°
    N.L.
Rubus occidentalis L. copses VI. 5 N.E.—All.—R.Mts.—Or.
Rubus villosus Ait. copses VIII. 6 Atl.—Up.Mo. G.—Can.
Rosa setigera Michx. bottom VI. 5 W.N.Y.—Fla.—Miss.—47°N.L.
Rosa carolina L. woods I. 2 Atl.—Miss. G.—Can.
Rosa parviflora Ehrh hillsides V. 5 N.F.—Fla.—Miss.
Rosa blanda Ait. prairies, hillsides III. 5 N.E.—Up.Miss. N. Mex.—Cal—
    69°N.L.
Crataegus coccinea L. wooded hillsides II. 3 N.E.—Up.Mo.—Fla.—N.Mex.
Crataegus tomentosa L. 5 bottom woods V. 4 N.E.—All.—La.—Up. Miss.
Crataegus subvillosa Schrad. bottom woods VI. 4 N.E.—La.—Up.Miss.
Crataegus crus galli L. bottom woods III. 5 Atl.—Miss. G.—Can.
Pyrus coronaria L. wooded hillsides, copses V. 5 W.N.Y.—All.—La.—Up.
    Miss.
Amelanchier canadensis T.Gr. var Botryapium wooded hillsides V. 3 Atl.—
    Cal. G.67°N.L.

Saxifragaceæ. In N. Am. 23 Gen. 135 Spec.

Ribes rotundifolium Michx. woods V. 5 N. E.—All. (N.Ca.)—La.—R.Mts.
Ribes floridum L. woods V. 3 N.Y.—Va.—Ky.—Up.Mo.—54°N.L.
Hydrangea arborescens L. wooded hillsides III. 4 N.J.—Fla.—Miss. G.—
    Can.
Parnassia caroliniana Michx. bottom, springy places III. 5 Atl.—Miss.G.—
    Can.
Saxifraga pennsylvanica L. wet prairies II. 2 N.E.—Va.—Up.Miss.
Heuchera hispida Ph. wooded hillsides V. 3 All. (Va.—N.Ca.)—Up.Miss.
Mitella diphylla L. rocky hillsides V. 6 N.E.—All. (N.Ca.)—Up.Miss.

Crassulaceæ. In N. Am. 5 Gen. 40 Spec.

Penthorum sedoides L. bottom V. 8 Atl.—Miss. G.—Can.

Hamamelaceæ. In N. Am. 15 Gen. 3 Spec.

Hamamelis virginica L. wooded hillsides II. 4 Atl.—Miss. G.—Can.

Haloragæ. In N. Am. 3 Gen. 14 Spec.

†Proserpinaca palustris L. swamps I. 3 Atl.—Up.Miss.—W.Tex. G.—Can.

Onagraceæ. In N. Am. 15 Gen. 148 Spec.
Circæa Lutetiana L. shady woods V. 3 N.E.—All.—La.—Up.Mo.
Gaura biennis L. fences, fields VIII. 5 N.Y.—Miss.—Ga. N.Mex.
Epilobium palustre L. var. lineare ditches V. 5 N.E.—All. (N.Ca.)—Up.Miss.
    Or.—Arct.
Epilobium coloratum Muhl. ditches, wet prairies III. 5 N.E.—All. (N.Ca.)
    Sonora—Cal.—Or.—54°N.L.
Oenothera biennis L. bottom, fields VI. 5 Atl.—W.Tex. Cal. Or. G.—56°N.L.
Oenothera rhombipetala Nutt. sandy soil I. 4 Up.Miss.—Tex.—Cal.
Oenothera fruticosa L. wet prairies II. 5 Conn.—All.—La.—Up.Mo.
†Ludwigia alternifolia L. fences, ditches I. 3 N.E.—Up.Mo. Fla.—La.
Ludwigia polycarpa Sh. Pet. bottom III. 3 Mich.—Oreg. Ky.—Sask.
Ludwigia palustris Ell. ditches, swampy places V. 8 Atl.—Pacif. G.—54°
    N.L.

Lythraceæ. In N. Am. 6 Gen. 13 Spec.
Rotala ramosior Koehne (Ammannia humilis Michx.) bottom IV. 3 Atl.—
    Miss.—Or. G.—Can.
Ammannia coccinea Rottb. (A. latifolia L. Mant.) bottom VI. 5 Oh.—La.—
    Up.Mo. Cal.—Sonora.
Lythrum alatum Ph. bottom VIII. 5 Mich.—Fla.—Sonora—Cal.
Cuphea viscosissima Jacq. dry wood openings VIII. 4 Conn.—All.—La.—
    Up.Miss.

Cucurbitaceæ. In N. Am. 11 Gen. 26 Spec.
Sicyos angulatus L. bottom V. 3 Atl.—Miss. G.—Can.
Echinocystis lobata T.Gr. bottom V. 8 W.N.E.—Ky.—Up.Mo.—Sask.

Ficoideæ. In N. Am. 3 Gen. 4 Spec.
Mollugo verticillata L. roadsides, waste places, yards VIII. 8 N.E.—Cal.
    Fla.—N.Mex.

Umbelliferæ. In N. Am. 45 Gen. 171 Spec.
Sanicula marilandica L. woods VI. 6 N.F. Ga.—La.—R.Mts.—Or.
Eryngium yuccaefolium Michx. dry prairie V. 5 N.J.—Up.Mo. Fla.—W.
    Tex.
Heracleum lanatum Michx. banks, copses III. 5 N.E.—All. (N.Ca.)—N.Mex.
    Cal.—Sitka—Huds.—58°N.L.
Archemora rigida D.C. springy places III. 4 N.Y.—Fla.—Miss.
Archangelica atropurpurea Hoffm. springy places II. 5 N.E.—Up.Miss.
Thaspium barbinode Nutt. banks I. 4 N.E.—Fla.—Up.Miss.
Thaspium aureum Nutt. woods V. 5 Atl.—Miss. G.—Can.
Thaspium trifoliatum Gr. woods V. 5 Atl.—Miss.—R.Mts. G.—Sask.
Pimpinella integerrima Benth Hook (Zizia D. C.) wooded hillsides V. 5
    N.E.—Up.Miss,—La.
Cicuta maculata L. swamps V. 5 Atl.—Pacif. G.—Sask.
Cicuta bulbifera L. swamps II. 5 N.E.—Up.Miss.
Sium cicutæfolium Gm. swamps V. 5 Atl.—Pacif. G.—Hudson's Bay.
Berula angustifolia Koch. springs, creeks II. 7 Mass.—Pacif. (Sonora.—
    Oreg.)
Cryptotænia Canadensis D.C. woods IV. 5 N.E.—All.—La.—Up.Mo.
Chærophyllum procumbens Lam. woods V. 5 N.J.—All.—La.—Up.Miss.
Osmorrhiza brevistylis D.C. woods II. 4 N.E.—N.Ca.—Cal.—Sitka.
Osmorrhiza longistylis D.C. woods II. 3 N.E.—Or. La.—Sask.

Araliaceæ. In N. Am. 2 Gen. 9 Spec.

Aralia racemosa L. woods V. 3 N.E.—All. (Ga.)—R.Mts.—Sask.
Aralia nudicaulis L. woods III. 3 Labrador—All. (N.Ca.)—R.Mts.—64°N.L.
Aralia quinquefolia Gr. woods III. 4 N.E.—All.—La.—Up.Miss.

Cornaceæ. In N. Am. 3 Gen. 20 Spec.

†Cornus circinata L. woods I. 1 Can.—All. (Va.)—Ill.
Cornus sericea L. woods, copses III. 4 Atl.—Miss. G.—Can.
Cornus stolonifera Michx. bottom woods II. 3 N.E.—N.Mex.—Alaska—Huds.—69°N.L.
Cornus asperifolia Michx. bottom woods V. 5 Fla.—S.Ca. Ky. Ill.—La.
Cornus paniculata L'Her. woods V. 3 N.E.—Up.Miss. N.Ca.—La.
Cornus alternifolia L. woods IV. 3 N.E.—Fla.—Up.Miss.

Caprifoliaceæ. In N. Am. 8 Gen. 47 Spec.

Lonicera flava L. rocky banks II. 3 N.Y.—All. (Ga.)—Up.Miss.
†Lonicera parviflora Lam. rocky banks I. 1 N.E.—All. (N.Ca.)—R.Mts.—Huds.
Triosteum perfoliatum L. woods IV. 4 N.E.—All. (Ga.)—Up.Miss.
Sambucus canadensis L. bottom VII. 5 Atl.—R.Mts. G.—Sask.
Viburnum Lentago L. banks, hillsides III. 3 N.E,—All. (Ga.)—Miss.—Sask.
Viburnum prunifolium L. copses III. 3 Conn.—Up Miss. G.—Can.
Viburnum dentatum L. wooded hillsides I. 2 Vt,—N.J.—Ky.—Wisc.
Viburnum opulus L. wooded hillsides I. 2 Vt.—R,Mts.—Arct.

Rubiaceæ. In N. Am. 20 Gen. 82 Spec.

Galium Aparine L, bottom, copses IV. 5 N.E.—Sonora—Alaska.
Galium concinnum T.Gr. dry woods V. 8 All. (Pa.—Va.)—Up.Miss.
Galium trifidum L. moist woods V. 8 Atl.—Pacif.—Alaska G.—68°N.L.
Galium triflorum Michx. shady woods IV. 5 Atl.—Pacif. G.—Greenland.
Galium circæzans Michx. shady woods V. 5 Atl.—Miss. G.—Can.
†Spermacoce glabra Michx. banks I. 1 Fla.—Tex.—Ills.—Oh.
Cephalanthus occidentalis L. bottom, swamps V. 8 Atl.—Tex.—Cal. G.—Can.

Compositæ. In Am. 221 Gen. 1557 Spec.

Vernoniaceæ.
Vernonia fasciculata Michx. bottom VIII. 8 Oh. Ky.—Up.Mo. Fla.—La.
Eupatoriaceæ.
Liatris cylindracea Michx. dry woods and prairies, IV. 5 W.Can.—Up. Miss.—La.
Liatris scariosa Willd. dry sandy soil V. 6 W.Can.—Fla.—Tex.—R.Mts.—Sask.
Liatris pycnostachya Michx. dry prairie V. 5 Ill.—La.—Tex.
Kuhnia eupatorioides L. dry woods and prairies VIII 8 N.J.—R.Mts. Fla.—N.Mex.
Eupatorium purpureum L. bottom, copses VI. 6 Atl.—Up.Mo. R.Mts. G.—Can.
Eupatorium altissimum L. dry copses III. 4 All.—La.—Up.Mo.
Eupatorium sessilifolium L. copses II. 3 N.E.—All. (Ga.)—Up.Miss.
Eupatorium perfoliatum L. bottom V. 5 Atl.—Up.Mo. G.—Can.
Eupatorium serotinum Michx. bottom VII. 9 N.Ca.—Up.Mo. Fla.—Tex.
Eupatorium ageratoides L. shady wooded hillsides VI. 6 Atl.—Up.Mo. G.—Can.

Asteroideæ.

Aster corymbosus Ait. woods II. 3 N.E.—All. (Ga.).—Up.Miss.

Aster sericeus Vent. dry, gravelly hillsides VI. 6 Oh.—Up.Mo.—All. (N. Ca.)—La.—Tex.

Aster lævis L. woods and dry hillsides V. 5 N.E.—All.—La.—N.Mx.—Sask.

Aster azureus Lindl. copses, dry prairies V. 4 Oh. Micb.—Up.Mo. All. (Ga.) —La.

Aster Shortii Boot wood VI. 6 Oh.—Up. Miss. All. (Ga.)—La.

Aster Drummondii Lindl. woods, copses VI. 6 Oh.— Miss. Ga.—Can.

Aster cordifolius L. woods VI. 6 N. E.—La.—Up.Miss.

Aster sagittifolius Willd. woods VI. 6 Atl.—Up.Mo. W.Tex. G.—Can.

Aster ericoides L. sandy soil, prairies II. 4 Conn.—Wisc. Fla.—La.—R.Mts.

Aster multiflorus Ait. sandy soil, prairies VII. 8 Atl.—Mex.—Alaska G.— Arct.

Aster Tradescanti L. fields, banks, etc. V. 3 N.E.—La.—Up.Miss.

Aster diffusus Ait.* fields, copses X. 10 Atl.—Up.Mo. G.—Can.

Aster paniculatus Lam. bottom VI. 8 Atl.—N.Mex.—Or. G.—Can.

Aster salicifolius Ait. bottom II. 4 N.E.—W.Tex.—Can.

Aster junceus Ait. bottom II. 4 Oh.—Cal.—Arct.

Aster Novi-Belgii L. bottom III. 8 Atl.—N.Mex.—Or.

Aster puniceus L. bottom V. 8 N.E.—Miss. Fla.—N.Mex. G.—Huds.

Aster prenanthoides Muhl. bottom I. 3 N.Y.—All. (N.Ca.)—Up.Miss.

Aster oblongifolius Nutt. dry gravelly hillsides III. 5 Pa. Va.—N.Mex.

Aster amethystinus Nutt. bottom I. 2 Mass.—Ill. Wisc.

Aster Novæ Angliæ L. copses, fences VI. 5 N.E.—All. (Ga.)—Up.Miss.— R.Mts.

Aster anomalus Engelm. wood openings III. 4 Up.Miss.

Aster umbellatus Mill. (Diplopappus umbellatus and amygdalinus T.Gr.) springy places in woods II. 4 N.F.—All.—Up.Miss.

Aster infirmus Michx. (Diplopappus cornifolius T.Gr.) moist woods I. 3 N.E.—All.—La.—Up.Miss.

Aster linariifolius L. (Diplopappus linariifolius Hook.) sandy hills II. 5 Atl.—Miss. G.—Can.

Erigeron canadensis L. waste places X. 10 Atl.—Pacif. G.—Sask.

Erigeron divaricatus Michx. sandy soil II. 5 Ky. Ill.—La.—W.Tex.

Erigeron bellidifolius Muhl. copses, hillsides, V. 5 N.E.—All.—La.—64° N.L.

Erigeron philadelphicus L. banks, moist places, V. 5 Atl.—Pacif. G.—Arct.

Erigeron annuus Pers. fields, waste places VII. 5 N.E.—Ky.—Up.Mo.

Erigeron strigosus Muhl. fields, wood openings VI. 5 Atl.—Pacif. G.—Sask.

Boltonia asteroides L'Her. (including B. glastifolia L'Her.) bottom V. 7 W. Can.—Up.Miss.—Fla.—La.

Solidago latifolia L. shady woods V. 5 N.E.—All. (Ga.)—Up. Miss.

Solidago speciosa Nutt. copses V. 4 Atl.—Tex. Up.Mo. G.—Can.

Solidago rigida L. dry hillsides VI. 5 Conn. All. (N.Ca.)—Tex.—R.Mts.— Sask.

Solidago Ohioensis Ridd. swamps II. 5 W.N.Y.—Ill.

Solidago Ridellii Frank swamps III. 5 Oh.—Up.Miss.

Solidago neglecta T.Gr. swamps II. 5 N.E.—Up.Miss.

Solidago patula Muhl. springy places III. 4 Atl.—Miss. G.—Can.

---

*There is no genus, the species of which are more subject to changes in nomenclature and so difficult to limitate, than the genus Aster, particularly the sections dumosi and salicifolii of Torr. & Gr. Gray in his new " Synoptical Flora " of North America does unite A. simplex and A. tinuifolius Nees (partim) under the name A. paniculatus Lam. A. carneus is partly put under A. salicifolius Nees, A. æstious Gr. under A. junceus Ait., A. miser L ? is A. diffusus Ait. The genus Diplopappus is re-united with Aster.

Solidago arguta Ait. copses V. 5 N.E.—All. (S.Ca.)—Miss.—Huds.
Solidago ulmifolia Muhl. woods VI. 6 N.Y.—All.—Up.Mo.
Solidago nemoralis Ait. dry wood openings VI. 6 Atl.—R.Mts. W. Tex. G.—
  Sask.
Solidago missouriensis Nutt. dry praries III. 5 Up.Miss.—R.Mts. La.—As-
  siniboin.
Solidago canadensis L. copses, fence rows X. 8 Atl.—N.Mex. G.—Subaret.
Solidago serotina Ait. var. gigantea banks, copses III. 5 Atl.—Or. G.—Sask.
Solidago tenuifolia Ph banks II. 3 Atl.—Up.Mo. G.—Can.
Solidago lanceolota L. banks, bottom V. 5 Atl.—R.Mts. G.—Huds.
Chrysopsis villosa Nutt. dry prairie III. 5 Ill. Ky. Ala.—N.Mex.—Or.—
  Sask.

· Senecioideæ.

Polymnia canadensis L. woods III. 3 Can.—All.(N.Ca.)—Up.Miss.
Silphium laciniatum L. dry prarie V. 5 Oh.—Up. Mo. Ala.—Tex.
Silphium terebinthinaceum L. dry prairie V. 5 Mich. Oh. Up.Miss.—W. Ga.
Silphium integrifolium Michx. dry prairie III. 3 W.Ga.—Up.Mo.
Silphium perfoliatum L. copses III. 3 Mich.—Up.Mo.—La.—Ga.
Parthenium integrifolium L. dry prairie, wood openings III. 2 All.—Up.
  Miss.—Tex.
Ambrosia trifida L. bottom, fence-rows, fields III. 7 Atl.—Up.Mo. G.—Can.
Ambrosia artemisiæfolia L. bottom, waste places X. 10 Atl.—Pacif. G.—
  Sask.
Ambrosia psilostachya D.C. sandy soil II. 5 Ill.—Tex.—Cal.
†Ambrosia bidentata Michx. prairie I. 1 Up.-Miss.—La.-Tex.
Xanthium canadense Mill. bottom X. 10 Up.Miss.—Sask.—Cal.—Tex.
Eclipta alba Hassk bottom III. 5 Pa.—Up.Mo.—Or.—Fla.—Tex.
Heliopsis lævis Pers. copses IV. 5 Atl.—Up.Mo.—N.Mex. G.—Sask.
Echinacea purpurea Moench. copses IV. 4 All.—Up.Mo.
Eschinacea angustifolia D.C. prairies V. 6 Ill.—Up.Mo.Ala.—Tex.
Rudbeckia laciniata L. bottom VIII. 8 Atl.—N.Mex.G.—Can.
Rudbeckia subtomentosa Ph. prairie III. 5 Up.Miss.—La.
Rudbeckia triloba L. woods V. 5 Pa.—W.Fla.—Miss.
Rudbeckia hirta L. dry prairie, hillsides V. 7 W.N.Y.—Up.Miss. Fla.—La.
Lepachys pinnata T.Gr. prairie, hillsides V. 4 W.N.Y.—Up.Miss. Fla.—La.
Helianthus rigidus Desf. prairie V. 5 Mich. W.Ga.—N.Mex.—Sask.
Helianthus laetiflorus Pers. prairie I. 4 Oh. W.Ga.—Up.Miss.
Helianthus occidentalis Ridd. prairie V. 5 Oh.—All. (N.Ca.)—Miss.
Helianthus giganteus L. copses, bottom II. 5 N.E.—All.—Miss.—Sask.
Helianthus grosseserratus Martens copses, fields, VIII. 7 Oh.—Up. Miss.
  La.—N.Mex.
Helianthus strumosus L. banks copses VII. 5 Atl.—Up.Mo.G.—Can.
Helianthus tracheliifolius Wild. copses II. 4 Pa.—Up.Mo.
Helianthus divaricatus L. copses V. 5.—Fla.—La.—Can. Sask.
Helianthus decapetalus L. copses III. 5 N.E.—All.(Ga.)—Miss.
Helianthus doronicoides Lam. copses fence rows V. 5 All.—Up.Mo.
Helianthus hirsutus Raf. copses I. 2 Oh.—Ga.—Tex. Wisc.
Verbesina helianthoides Benth. Hook. copses I. 2 Oh.—W.Ga.—Up.Mo.
Actinomeris squarrosa Nutt. copses V. 5 All.—Up.Mo.
Coreopsis aristosa Michx. swamps VIII. 8 Oh.—Wisc.—La.
Coreopsis tripteris L. copses, banks V. 3 Pa.—Fla.—La.—Up.Miss.
Coreopsis palmata Nutt. prairie, hillsides V. 4 La. W.Tex.—Wisc. Lake
  Winnipeg.

Coreopsis lanceolata L. prairie II. 4 W.Can.—Ill. Fla.—La.
Bidens frondosa L. waste places; bottom, VIII. 5 Atl.—Up.Mo. G.—Can.
—Or.
Bidens connata Muhl. low banks V. 5 N.E.—All. (Ga.)—Up.Mo.
Bidens chrysanthemoides Michx. swamps, banks VIII. 8 Atl.—Pacif.Gult
Can.
Dysodia chrysanthemoides Lag. waste places, roadsides VI. 8 Up.Miss.—
La. Up.Mo.—Mex.
Helenium auctumnale L. bottom X. 6 Atl.—N.Mex.—Ore. G.—Subarct.
Achillea millefolium L. prairies, open roads V. 3 Atl.—Pacif. G.—Arct.
Artemisia caudata Michx. prairie IV. 5 N.E.—Up.Mo. Sask.—Tex.
Artemisia ludoviciana Nutt. banks II. 2 Pacif.—Ill.—Sask.
Artemisia biennis Willd. wastes places IV. 6 Pacif.—Tenn.—Mackenzie R.
Gnaphalium polycephalum Michx. dry prairie, wood openings VI. 5 Atl.—
Mex. G.—Can.
†Gnaphalium purpureum L. sandy soil I. 1 Atl.—Pacif. G.—Can.
Antennaria plantaginifolia Hook. open woods, hillsides VI. 5 Atl.—Pacif.
G.—Huds.
Erechtites hieracifolia Raf. waste places VI. 4 Atl.—Up.Mo. G.—Sask.
Cacalia suaveolens L. bottom I. 3 Conn.—Ill. All.—W.Fla.
Cacalia reniformis Muhl. woods II. 3 Pa.—N.Ca.—Up.Miss.
Cacalia atriplicifolia L. bottom V. 5 W. Can.—Ill.—Fla.
Cacalia tuberosa Nutt. prairie II. 3 W.Can.—Ala. Wisc.
Senecio aureus L. wood, hillsides V. 5 N.F—Fla.—Tex.—Or.

Cynareæ.

Cnicus altissimus Willd. copses, fields IV. 3 N.Y.—Fla.—Tex.—Wisc.
Cnicus altissimus var discolor Gr. copses, roadsides V. 4. Can.—Ga.—Ill.
Cnicus muticus Ph. bottom, springy places III. 3 N.F.—Fla.—La.—Sask.
Cnicus pumilus Torr. prairie V. 4 N.E.—U.Miss.

Cichoriaceæ.

Krigia amplexicaulis Nutt.(Cynthia virginica Don) open woods V. 3 N.Y.—
Colorado. Ga.—Minnesota.
Hieracium scabrum Michx. open woods III. 3 N.E.Can.—Ga.—Up.Miss.
Hieracium longipilum Torr. prairies V. 3 Mich.—Ark.—W.Tex.—Up.Mo.
Prenanthes alba L. (Nabalus Hook) shady hillsides V. 4 N.F.—All.—U.Miss.
Prenanthes aspera Michx. dry prairie III. 3 Oh.—La.—Up. Miss.
Prenanthes crepidinea Michx. bottom, rich soil II. 3 W.N.Y.—Ill.—Ky.
Prenanthes racemosa Michx. wet prairie III. 3 N.E.Can.—Up.Miss.—Sask.
R.Mts.
Troximon cuspidatum Ph. prairie, hillsides II. 3 Ill.Wisc.—R.Mts.
Taraxacum officinale Weber open wood, yards V. 7 N.E.—R.Mts. Ore.—
Arct.
Lactuca canadensis L. copses, roadsides V. 3 Atl.—N.Mex. G.—Sask.
Lactuca integrifolia Big. open ground II. 3 N.E—Ga.—Ill.
Lactuca hirsuta Muhl. wooded hillsides III. 4 N.E—Up.Miss.—Tex.
Lactuca floridana Gaertn. (Mulgedium DC.) copses, wooded hillsides V. 3
Pa.—Ill. Fla.—Tex.
Lactuca acuminata Gr. copses, wooded hillsides II. 2 N.Y.—Fla.—Ill.

Lobeliaceæ. In N. Am. 5 Gen. 29 Spec.

Lobelia cardinalis* L. bottom IV. 6 Atl.—Up.Mo. G.—Can.
Lobelia syphilitica L. bottom VI. 4 N.E.—Atl.—La.—R.Mts.

*A single hybrid of Lobelia cardinalis and syphilitica was found.

Lobelia leptostachys A. D. C. dry open woods V. 4 Ill.—Up.Mo. S.Ca.—Tex.
Lobelia inflata L. woods V. 4 N,E.—All.—La.—Up.Miss.
Lobelia spicata Lam. dry prairie, open woods II. 3 N.E.—All.—La. Up.Mo.
Lobelia kalmii L. swamps, wet banks III. 7 N.E.—La.—Ill.

Campanulaceæ. In N. Am. 4 Gen. 18 Spec.
Campanula aparinoides Ph. moist grassy plains III. 5 N.E.—La.—Up.Miss.
Campanula americana L. bottom, copses V. 4 Atl.—Miss. G.—Can.
Specularia perfoliata A.D.C. dry hillsides VI. 6 Atl.—Pacif. G.—Can.

Ericaceæ. In N. Am. 34 Gen. 131 Spec.
Vaccinium vacillans Sol. wooded hillsides I. 6 N.E.—S.Ca.—Ga.
Gaylussacia resinosa** T. Gr. wood I. 2 N.E.—Atl. (Ga.)—Up. Miss.
Arctostaphylos uva ursi Spr. wooded hillsides I. 4 N.J.—R.Mts. Cal.—Arct.
Monotropa uniflora L. woods III. 2 Atl.—Or. G.—Can.

Ebenaceæ. In N. Am. 1 Gen. 2 Spec.
Diospyros virginiana L. woods along river banks II. 4 R.I.—Kas.—Fla.—
Tex.

Primulaceæ. In N. Am. 11 Gen. 36 Spec.
Androsace occidentalis Ph. sandy hillsides, banks VI. 6 Ill.—Ark.—R.Mts.
—N.Mex.
Dodecatheon meadia L. open woods II. 3 Pa.—N.Ca.—La.—Cal.—Alaska.
Steironema ciliatum Raf. (Lysimachia L.) wood bottom V. 5 N.E.—La.—
Up.Mo.
Steironema lanceolatum Gr. bottom V. 6 N.E.—Up.Miss.
Steironema longifolium Gr. wet prairie III. 4 Pa.—S.Ca.—Up.Miss.
Lysimachia thyrsiflora L. swamps II. 3 Pa.—Up.Miss.—Subarct.
Samolus valerandi L. var. americanus Gr. bottom IV. 5 Atl.—Mex.—Or. G.—
Can.

Oleaceæ. In N. Am. 6 Gen. 29 Spec.
Fraxinus americana L. bottom, hillsides IV. 4 N.E.—Up.Mo. Sask —Fla.—
Tex.
Fraxinus pubescens Lam. bottoms II. 2 N.Br.—Minn. Fla.—Ala.
Fraxinus viridis Michx. f. bottom hillsides VI. 5 N.E.—Sask. Fla.—Ariz.
Fraxinus quadrangulata Michx. bottom III. 3 Mich.—Minn.—Ark.
Fraxinus sambucifolia Lam. bottom III. 4 N.F.—Sask. Va.—Ark.

Apocynaceæ. In N. Am. 9 Gen. 21 Spec.
Amsonia tabernaemontana Walt. bottom VI. 4 Ill.—Tex.—Fla.—N.Ca.
Apocynum androsaemifolium L. copses, hillsides IV. 3 Can.—Ga.—N.Mex.
—Br.Columb.
Apocynum cannabinum L. copses, banks V. 5 Atl.—Pacif. G.—Can.

Asclepiadaceæ. In N. Am. 17 Gen. 87 Spec.
Asclepias tuberosa L. dry open woods, hillsides, prairie VI. 5 Atl.—Mex.—
R.Mts. G.—Can.
Asclepias purpurascens L. copses, hillsides V. 3 N.E.—Up.Miss.—Tenn.
Asclepias incarnata L. wet bottom, banks VI. 6 N.E.—All. (Ga.)—Up.Miss.
  —N.Mex.
Asclepias cornuti Desn. fields, hillsides V. 5 Can.—Sask.—N.Ca.
Asclepias Sullivanti Engelm. wet prairie II. 3 Oh.—N.Mex.
Asclepias obtusifolia Michx. dry prairie IV. 3 N.E.—Up.Mo.—Fla.—Tex.
Asclepias Meadii Torr. dry prairie II. 3 Up.Miss.

**Gaylussacia was found about fifteen miles southwest and may occur somewhat nearer.

Asclepias phytolaccoides Ph. shady banks IV. 3 N.E.—All.(Ga.)—Up.Miss.
—La.

Asclepias quadrifolia L. wooded hillsides III. 3 Can.—N.Ca.—Up.Miss.—
Ark.

Asclepias verticillata L. dry prairie V.S. Atl.—R.Mts.—Mex.G.—Can.

Acerates longifolia Ell. prairie II 2 Fla.—Tex. Oh.—Up. Miss.

Acerates viridiflora Ell. dry prairie hillsides V. 4 N.E.—Sask. Fla.—Tex.

Gentianaceæ. In N.Am. 13 Gen. 82 Spec.

Gentiana quinqueflora Lam. shady hillsides IV. 4 N.E.—Up.Miss.—Fla.—
Tex.

Gentiana alba Muhl. copses III. 3 N.Y.—All. (Va.) Ky.—Up.Miss.

Gentiana Andrewsii Gris. bottom III. 4 N.E.—Sask.—All. (Ga.)

Gentiana puberula Michx. dry prairie, hillsides, II. 4 Oh. Ky.—Up.Miss.

†Menyanthes trifoliata L. swamps, I. 4 N.E.—Cal.—Alaska—Arct.

Polemoniaceæ. In N.Am. 5 Gen. 123 Spec.

Phlox maculata L. wooded hillsides I 2 Pa.—Up.Miss.—Fla.

Phlox glaberrima L. bottom, wet prairie II. 5 All. (Va.)—Up.Miss.—Fla.—
La.

Phlox pilosa L. prairie, hillsides V. 5 N.J.—Sask. Fla.—Tex.

Phlox divaricata L. woods VII. 7 N.Y.—Up.Miss. Fla.—Ark.

Phlox bifida Beck sandy soil, hillsides II. 5 Up.Miss.

Polemonium reptans L. wooded hillsides VI. 5 N.Y.—Ala.—Up.Miss.—Or.?

Hydrophyllaceæ. In N.Am. 14 Gen. 115 Spec.

Hydrophyllum virginicum L. woods V. 3 N.E.—All. (S.Ca.)—W.T.—Alaska.

Hydrophyllum appendiculatum Michx. woods IV. 4 Can.—Wisc. All.—Mo.

Ellisia nyctelea L. shady woods VI. 8 Pa.—Va.—La.—Sask.

Borragiuaceæ. In N.Am. 19 Gen. 123 Spec.

Echinospermum virginicum Lehm. (Cynoglossum Morisoni DC.) woods
VIII. 8 Can.—Alab.—Miss.

Mertensia virginica DC. bottom, shady hillsides V. 8 N.Y.—All.(S.Ca.)—
Up.Miss.

Myosotis verna Nutt. dry soil II. 4 N.E.—Fla.—Tex.—Cal. Or.

Lithospermum canescens Lehm. prairie V. 3 Can.—Sask. Ala.—Ariz.

Lithospermum hirtum Lehm. dry open wood, prairie V. 5 Va.—Mich. Up.
Miss. Fla.—Tex.

Lithospermum angustifolium Michx. (L. longiflorum Lehm.) prairie III. 6
Up.Miss.—Sask.—Ariz.

Lithospermum latifolium Michx. woods V. 3 Can.—Va.—Tenn. Up.Miss.

Onosmodium carolinianum DC. var. molle dry prairie, roadsides V. 6 Ill.—
Sask.—Tex.

Convolvulaceæ. In N.Am. 8 Gen. 73 Spec.

Jpomoea pandurata Mey. bottom, fields V. 6 Can.—Fla.—Up.Miss. Tex.

Jpomoea lacunosa L. bottom IV. 5 Pa.—Ill. S.Ca.—Tex.

Convolvulus sepium L. bottom, copses III. 3 N.E.—Fla.—N.Mex.—Sask.

Convolvulus spithameus L. dry open woods IV. 3 Can.—Fla.—Up.Miss.

Cuscuta tenuiflora Engelm. bottom upon Cephalanthus V. 5 Pa.—Sask.—
Ariz.

Cuscuta inflexa Engelm. bottom on shrubs V. 5 N.E.—Neb.—Ark.

Cuscuta chlorocarpa Engelm. bottom on Polygonum, etc. II. 3 Pa.—Up.
Miss.—Ark.

Cuscuta Gronovii Willd. bottom on Saururus, etc. V. 8 Can.—Up.Miss. Fla.
—Tex.

Cuscuta compacta Juss. bottom on shrubs V. 5 Can.—All. (Ala.) Up.Miss.
—Tex.
Cuscuta glomerata Chois. bottom on compositae V. 5 Oh.—Up.Miss.—Tex.

Solanaceæ. In N. Am. 13 Gen. 68 Spec.

Solanum nigrum L. roadsides, fields, woods V. 3 Atl.—Pacif. G.—Can.
Solanum carolinense L. sandy soil, roadsides, fields VI. 5 Conn.—Up.Miss.
Fla.—Tex.
Physalis virginiana Mill. (Ph. viscosa Gr.) sandy soil, fields, roadsides V. 4
Can.—Fla.—Tex.—Up.Miss.
Physalis lanceolata Michx. (Ph. pennsylvanica Gr.) sandy fields III. 4 Fla.
N.Mex.—Sask.
Datura tatula L. waste places, roadsides N. 8 Atl.—Pacif. G.—Can.

Scrophulariaceæ. In N. Am. 37 Gen. 313 Spec.

Scrophularia nodosa L. banks, copses VI. 4 Atl.—Pacif. S.—Can.
Chelone glabra L. bottom, springy places IV. 4 N.F.—Fla.—Ark.—Sask.
Pentstemon pubescens Sol. dry prairie, open woods VIII. 5 Can. Up.Miss.
Fla.—Tex.
Mimulus ringens L. bottom VI. 5 N.E.—Up.Miss.—Tex.
Mimulus Jamesii T. Gr. springs I. 5 Mich. Up.Miss.—R.Mts.—Mex.
Conobea multifida Benth. sandy banks VI. 6 Oh.—Ill.—Tex.
Gratiola virginiana L. bottom VI. 5 Atl.—Tex.—Or.
Ilysanthes gratioloides Benth. bottom VI. 5 Atl.—Tex.—Or.
Veronica virginica L. woods V. 4 Can.—Sask. Ala.—Up.Mo.
Veronica anagallis L. ditches, brooks II. 5 Can.—Up.Miss. N.Mex.—Br.Col.
Veronica americana Schwein ditches, brooks I. 4 N.E.—Ark.—Up.Miss.
N.Mex. Cal.—Alaska.
Veronica scutellata L. bottom I. 5 N.E.—Up.Miss. Huds.—Br.Col. Cal.
Veronica peregrina L. wet fields, bottom X. 10 Atl.—Pacif. G.—Huds.
Seymeria macrophylla Nutt. bottom IV. 4 Oh.—Ky.—La.—Tex.
Gerardia purpurea L. bottom IV. 5 Can.—Fla.—Tex.—Up.Miss.
Gerardia tenuifolia Vahl. bottom, open woods V. 6 Atl.—R.Mts. G.—Can.
†Gerardia aspera Dougl. bottom I. 3 Up.Miss.—Sask.—W.Ark.
Gerardia grandiflora Benth. woods, copses IV. 5 Up.Miss.—Tenn.—Tex.
Gerardia auriculata Michx. fields, bottom II. 5 All. (Pa.—N.Ca.)—Up.Miss.
Ark.
Castilleja coccinea Spr. dry open woods, copses VI. 6 All. Can.—Sask.—Tex.
Pedicularis canadensis L. prairie, hillsides V. 5 Can.—Sask. Fla. R.Mts.—
Mex.
Pedicularis lanceolata Ph. swamps, springy places IV. 7 Conn.—Va.—Sask.

Orobanchaceæ. In N. Am. 4 Gen. 13 Spec.

Aphyllon uniflorum Gr. wooded hillsides II. 2 N.F.—Tex. Cal. Br.Columbia.

Lentibulariaceæ. In N. Am. 2 Gen. 19 Spec.

Utricularia vulgaris L. waters III. 5 N.F.—Sask.—Br.Col. All.—Tex.
†Utricularia intermedia Hayne shallow water I. 5 N.F.—Up.Miss.—60°N.L.
Cal.

Bignoniaceæ. In N. Am. 4 Gen. 6 Spec.

Tecoma radicans Juss. bottom, woods III. 4 Pa.—Up.Miss. Fla.—Tex.

Acanthaceæ. In N.Am. 15 Gen. 39 Spec.

Ruellia ciliosa Ph. dry prairie VI. 5 Mich.—Up.Miss. Fla.—La.
Ruellia strepens L. woods II. 4 Pa.—Up.Miss. Fla.—Tex.

Verbenaceæ. In N.Am. 11 Gen. 35 Spec.

Phryma leptostachya L. woods VI. 3 Atl.—Miss. Ga.—Can.

*Verbena urticifolia L. bottom, roadsides VII. 7 Atl.—Mex.G.—Can.

Verbena hastata L. bottom, roadsides V. 5 Can.—Sask. Fla.—N.Mex.—Cal.

Verbena stricta Vent. dry prairie, roadsides × 8 Oh.—Up.Mo.—Tex. N.Mex

Verbena bracteosa Michx. roadsides VII. 7 Wisc.—W.Fla.—Ariz.—Or.

Lippia lanceolata Michx. bottom, banks VI. 8 Pa.—Up.Miss. Fla.—Mex.—
    Cal.

Labiatæ. In N.Am. 37 Gen. 196 Spec.

Isanthus cœruleus Michx. sandy banks, hillsides IV. 7 Can.—Ga.—Tex, |
    Up.Miss.

Teucrium canadense L. bottom VI. 5 Atl.—Mex.G.—Can.

Mentha canadensis L. bottom, banks IV. 5 Atl.—Pacif. Can.—Sask.

Lycopus virginicus L. bottom IV. 5 Labr.—Fla.—Br.Columb.

Lycopus sinuatus Ell. bottom III. 4 Can.—Fla.—Tex.--Or.

Pycnanthemum muticum Pers. var. pilosum Gr. dry open woods V. 5 Oh.--
    Ill. Ark.

Pycnanthemum linifolium Ph dry open woods II. 3 N.E.—Up.Miss. Fla.--
    Tex.

Pycnanthemum lanceolatum Ph. dry open woods, copses IV. 4 N.E.—Nebr.
    —Ga.

Hedeoma pulegioides Pers. open woods V. 8 N.E.—All. (S.Ca.)--Up.Miss.

Monarda fistulosa L. open woods, copses, hillsides V. 5 Can.--Fla. Ariz.--
    Br.Columb.

Monarda clinopodia L. shady woods II. 3 Can.—All. (Ga.)—Up.Miss.

Blephilia hirsuta Benth. woods V. 5 Can.—All. (Ga.)—Up.Miss.—Tex.

Lophanthus nepetoides Benth. copses IV. 3 Vt.—Wisc. All. (Ca.)—Tex.

Lophanthus scrophulariæfolius Benth. copses IV. 4 N.Y.—Up.Miss. N.Ca.
    —Ky.

Scutellaria lateriflora L. bottom VI. 6 Can—Fla.—N.Mex.—Br.Columb.

Scutellaria versicolor Nutt. open woods V. 3 Pa.—Up.Miss. Fla.—Tex.

Scutellaria parvula Michx. gravelly banks, hillsides III. 6 N.E.—Up.Miss.
    Fla.—Tex.

Scutellaria galericulata L. bottom II. 3 N.F.—All. (N.Ca.)—Cal.—60° N.L.

Scutellaria nervosa Ph. bottom I. 4 N.Y.—Va.—Up.Miss.

Brunella vulgaris L. bottom, woods VI. 4 N.F.—Fla.—Cal,—Alaska.

Physostegia virginiana Benth. bottom VI, 8 N.E.—Fla.—Tex.—65° N.L.

Stachys palustris L. bottom III. 3 N.F.—Pa.—R.Mts.

Stachys aspera Michx. bottom V. 5 Can.—Fla.—La.—Up.Miss.

Plantaginaceæ. In N.Am. 1 Gen. 14 Spec.

**Plantago Rugelii Decaisne fields, roadsides, woods VI. 8 Can. Vt·—Up.Miss.
    Ga.—Tex.

Plantago cordata Lam. banks of brooks I. 3 N.Y.—Up.Miss. Ala.—La.

Plantago virginica L. gravelly hillsides III. 5 N.E.—Up.Miss. Fla.—Tex.

Aristolochiaceæ. In N. Am. 2 Gen. 11 Spec.

Asarum canadense L. shady hillsides V. 4 N.E.—All. (N.Ca.)—Up.Mo.

Aristolochia serpentaria L. woods II. 2 Conn.—Fla.—Miss.

°Several hybrids of Verbena occur: V. stricta × urticifolia, V. stricta × bracteosa, V. hastata × urticifolia.

**Plantago Rugelii was formerly taken for Pl. major an immigrated plant, which, so far, was not yet observed in our vicinity. Our species, which is indigenous, differs from P. major in the number of seeds (only 4 to 9) and the longer at the apex attenuated spike. Plantago sparsiflora Michx. probably does not grow in Illinois. A specimen from South Illinois I received under that name is nothing else than a depauperate Plantago Rugelii.

Nyctaginaceæ. In N. Am. 10 Gen. 50 Spec.
Oxybaphus nyctagineus Sweet fence rows, banks V. 5 Up.Miss.—La.—N.
Mex.—Up.Mo.

Phytolaccaceæ. In N. Am. 4 Gen. 5 Spec.
Phytolacca decandra L. bottom, banks V. 3 N.E—Up.Miss. Fla.—N.Mex.

Chenopodiaceæ. In N. Am. 17 Gen. 83 Spec.
Chenopodium album L. bottom, fence rows X. 8 Atl.—Pacif.G.—Gr. Bear
Lake.
Chenopodium hybridum L. bottom V. 5 Atl.—Pacif.G.—Sask.

Amarantaceæ. In N. Am. 16 Gen. 46 Spec.
Montelia tamariscina Gr. bottom VIII. 8 Vt.—Up.Mo.—W.Tex.
Amarantus albus L. waste places IV. 5 Atl—N.Mex. R.Mts.—Or.
Amarantus retroflexus L. bottom, waste places VIII. 8 Atl.—N.Mex.—
R.Mts.
Amarantus blitoides Wats. waste places V. 7;Mex.—R. Mts.—Up.Miss.

Polygonaceæ. In N. Am. 16 Gen. 107 Spec.
Polygonum pennsylvanicum L. moist places VI. 6 Atl.—Mex.G.—Can.
Polygonum incarnatum Ell. moist places III. 5 Atl.—Mex.G.—Can.
*Polygonum hydropiper L. moist places, ditches VI. 10 N.E.—Ariz.
Polygonum acre H.B.K. bottom V. 3 Atl.—Mex.—Cal.G.—Can.
Polygonum hydropiperoides Michx. moist places V. 6 Atl.—Mex.G.—Can.
Polygonum amphibium L. waters, sloughs, banks X. 8 Atl.—Mex.G.—Or.
Gr. Slave Lake.
Polygonum virginianum L. woods V. 3 N.E.—Fla.—Up.Mo.
Polygonum aviculare L. roadsides, yards X. 10 Atl.—Mex.G.—72°N.L.
Polygonum erectum L. yards, roadsides V. 5 Atl.—Can.—Or.
Polygonum ramosissimum Michx. bottom IV. 4 N.E.—Up.Mo.
Polygonum tenue Michx. dry soil, hillsides V. 5 N.E.—All.—La.—Sask.—Or.
Mex.
Polygonum sagittatum L. bottom II. 8 Atl.—Miss.G.—Can.
Polygonum dumetorum L. copses, banks V. 5 Atl.—Miss.—R.Mts.G.—Can.
Rumex orbiculatus Gr. bottom, swamps II. 4 N.E—Up.Miss.
Rumex britannica L. banks, moist places IV. 5 N.Y.—W.Tex.
Rumex verticillatus L. moist soil, banks V. 5 Atl.—Miss.G.—Can.

Lauraceæ. In N. Am. 5 Gen. 8 Spec.
Sassafras officinalis Nees. woods V. 4 Atl.—-Up.Mo. G.—Can.

Thymeleaceæ. In N. Am. 1 Gen. 2 Spec.
Dirca palustris L. bottom, springy places II. 3. Atl.—Cal.G.—Can.

Santalaceæ. In N. Am. 3 Gen. 6 Spec.
Comandra umbellata Nutt. dry hillsides, copses IV. 5 N.E.- All. (Ga.)—
Mex.G.—Can.

Saururaceæ. In N. Am. 2 Gen. 2 Spec.
Saururus cernuus L. swamps, springy places V. 6 Atl.—Miss. G.—Can.

Ceratophyllaceæ. In N. Am. 1 Gen. 2 Spec.
Ceratophyllum demersum L. waters X. 10 Atl.—Pacif. G.—62°N.L.

Callitrichaceæ. In N. Am. 1 Gen. 6 Spec.
Callitriche heterophylla Ph. waters III. 5 Atl.—Pacif. G.—71°N.L.

---

*Polygonum hydropiper may be indigenous as well as not. Porter, in Wheeler's Report, says:
"Introduced?" De Candolle has it not in the list of introduced plants.

8

Euphorbiaceæ. In N. Am. 17 Gen. 164 Spec.

Euphorbia maculata L. roadsides, waste places, fields X. 10 Atl.—Cal. G.—Can.

Euphorbia hypericifolia L. (Preslii Guss.) bottom fields VIII. 8 Atl.—N.Mex. G.—Can.

Euphorbia dentata Michx. bottom III. 4 Pa.—Up.Miss. La.—Sonora.

Euphorbia heterophylla L. banks, hillsides IV. 5 Up.Miss.—Mex.

Euphorbia corollata L. prairies, hillsides VII. 7 N.Y.—Up.Mo. Fla.—Mex.

Euphorbia commutata Engelm. sandy banks I. 3 Va.—Fla.—Up.Miss.

†Euphorbia obtusata Ph. woods I. 3 Vt.—S.Ca.—R.Mts.

Acalypha virginica L. bottom, woods VII. 8 Atl.—N.Mex. G.—Can.

Croton glandulosus* L. sandy soil I. 3 Va.—Up.Miss. La.—W.Tex.

Urticaceæ. In N. Am. 13 Gen. 24 Spec.

Ulmus fulva Michx. woods, bottom V. 3 Can.—N.Fla.—Tex.—Up.Mo.

Ulmus americana L. woods, bottom VIII. 8 N.F.—Fla.—R.Mts.—52°N.L.

Celtis occidentalis L. woods, bottom III. 5 Can.—Up.Mo. Fla.—Tex.

Morus rubra L. woods VI. 4 N.E.—Up Mo. Fla.—Tex.

Urtica gracilis Ait. bottom V. 4 N.E.—All.—N.Mex.—Cal. Or.

Laportea canadensis Gaud. bottom V. 6 Atl.—Up.Mo. G.—Can.

Pilea pumila Gr. shady woods V. 6 Atl.—Up.Mo. G.—Can.

Böhmeria cylindrica Willd. bottom III. 4 Atl.—W.Tex. G.—Can.

Parietaria pennsylvanica Muhl. shady woods III. ¡8 N.E.—All.—N.Mex.—Up.Mo.

Humulus lupulus L. copses. bottom III. 4 N.E.—All.—N.Mex. R. Mts.—Sask.

Platanaceæ. In N. Am. 1 Gen. 3 Spec.

Platanus occidentalis L. woods, bottom VI. 5 N.E.—Up.Miss. Fla.—Tex.

Juglandaceæ, In N. Am. 2 Gen. 12 Spec.

Juglans cinerea L. woods V. 4 N.Br.—Up-Miss. All. (Ga.)—Ark.

Juglans nigra L. woods V. 5 N.E.—Up.Miss. Fla.—Tex.

Carya olivae formis Nutt. woods, bottom III. 3 Up.Miss. W.Ky.—La.—Tex.

Carya alba Nutt. woods VIII. 8 Can.—Up.Miss. W.Fla.—Tex.

Carya tomentosa Nutt. woods VI. 5 Can.—Up.Miss. Fla.—Tex-

Carya sulcata Nutt. woods, bottom I. 3 Pa.—Up.Miss.—Ark.

Carya porcina** Nutt. woods I. 2 N.E.—Up.Miss. Fla.—Tex.

Carya amara Nutt, woods VI. 6 N.E.—Up.Miss. Fla.—Tex.

Cupuliferæ In N. Am. 7 Gen. 46 Spec.

Quercus alba L. woods X. 8 N.E.—Up.Miss. Fla.—Tex. G.—46° N.L.

Quercus macrocarpa Michx. woods, bottom V. 5 N.Br.—Up.Mo. N.Ca.—W.Tex.

Quercus bicolor Willd. woods, bottom II. 5 N.E.—All. (Ga.)—Ark. Up.Miss.

Quercus prinoides Willd.(Qu.prinus acuminata Michx.) woods IV.¦4 N.E.—Up.Miss. All.—W.Tex.

Quercus imbricaria Michx. woods IV. 4 Ta.—Up.Miss.—All.—(Ga.)

Quercus nigra L. woods I. 2 N.Y.—Up.Miss. Fla.—W.Tex.

Quercus coccinea Wang. woods VIII. 6 N.E.—Up.Miss. All.—Fla.

Quercus rubra L. woods VI. 5 N.Scot.—Up.Miss. Fla.—Tex.

---

*From the locality on which Croton glandulosus was found only a few years ago (railroad track) it may be concluded that it might be an immigrant from the southern part of Illinois.

**Carya porcina—No doubt single trees exist in the neighborhood, although I have not seen such; for amongst the nuts of the common hickory, brought to market. single nuts of the tree were observed.

Quercus Leana Nutt.* (coccinea × imbricaria) I. 1 Oh. Ill.
Corylus americana Walt. copses VIII. 10 Can.—Sask. Fla.—Tex.
Carpinus americana Michx. woods V. 3 N.Scot.—Up.Miss. Fla.—Tex.
Ostrya virginica Willd. woods V. 5 Can.—Up.Miss.—Winnipeg. Fla.—La.

Salicaceæ. In N. Am. 2 Gen. 68 Spec.

Salix candida Willd. swampy bottom I, 4. N.E.—Up.Miss.—Sask.
†Salix tristis Ait. dry hillsides I. 2. N.E.—All. (Ga.)—Up.Miss.
Salix humilis Marsh. dry hillsides V. 5. N.E.—All. (Ga.)—Up.Miss.
Salix discolor Muhl. banks bottom VII. 6. N.F.—N.Ca.—R.Mts.—Athabasca
   R.—Labr.
Salix sericea Marsh. (incl. S petiolata) banks bottom IV. 4. N.E.—Up.Miss.
   —Sask.
Salix cordata Muhl. var angustata bottom V. 5. N.E.—R.Mts.—Arct.
Salix nigra Marsh. bottom X. 8 N.Br.—Cal. Fla.—Ariz.
Salix amygdaloides Anders. bottom III. 4 N.Y.—Sask.—Or. Oh.—N.Mex.
Salix longifolia Muhl. bottom low banks VIII. 8 N.E.—La.—Ariz.—Cal.—
   Sask—66° N.L.
Salix myrtilloides L. swampy bottom I. 4. N.E.- Up.Miss.—Arct.
Populus tremuloides Michx woods III. 5 N.F.—N.Mex. Cal.—Alaska—Huds.
   69° N.L.
Populus grandidentata Michx. wooded hillsides V. 5 N.Br.—Up.Miss.—All.
   (N.Ca.)
Populus monilifera Ait. bottom X. 8 Vt.—Fla.—N.Mex.—R.Mts.

Coniferæ. In N. Am. 15 Gen. 82 Spec.

Juniperus virginiana L. hillsides II. 4 N.Br.—Up.Miss. Fla.—Tex.—R.Mts.
   67° N.L.
†Thuja occidentalis** L. swamps I. 1 N.Br.—All (N.Ca.)—Up.Miss.—Huds.

Araceæ. In N. Am. 9 Gen. 11 Spec.

Arisaema triphyllum Torr. woods V. 4 Atl.—Up.Mo. G.—Can.
Arisaema Dracontium Schott woods IV. 3 Atl.—Miss. G.—Can.
Peltandra virginica Raf. swampy bottom II. 3 Atl.—Miss. G.—Can.
Symplocarpus foetidus Salisb. swampy bottom III. 5. N.E.—N.Ca.—Up.
   Miss.
Acorus calamus L. swamp II, 6 Atl.—Miss. G.—Can.

Lemnaceæ. In N.Am. 2 Gen. 9 Spec.

Lemna trisulca L. waters III. 10 N.E.—N.Mex.—Cal.—55° N.L.
Lemna minor L. waters V. 10 N.E.—Fla.—Mex.—Or.—60° N.L.
Lemna polyrrhiza L. waters V. 10 Atl.—W.Tex. Nevada G.—55° N.L.
Wolffia columbiana Karsten waters I. 10 N.E.—La.—Up.Miss.

Typhaceæ. In N.Am. 2 Gen. 5 Spec.

Typha latifolia L. swamps, slough, banks III. 5 Atl.—Pacif. G.—60° N.L.
Sparganium eurycarpum Engelm. banks, ditches V. 5 N.E.—Pa.—Or.—Slave
   Lake.

Najadaceæ. In N.Am. 7 Gen. 29. Spec.

Najas flexilis Rostk. Illinois R. II. 8 N.E.—Up.Miss. Fla.—W.Tex.
Zannichellia palustris L. creeks I. 10 N.Y.— Sask.— Or. Cal. W. Fla. —
   Mex.

---

*The only tree unfortunately was cut last year: It stood on the bluff side near the western city limits.

**Thuja occidentalis is certainly extinct now in our flora, but in 1853 yet a large tree nearly two feet in diameter stood in a swamp at the foot of the eastern bluff. The locality and the age of the tree is against the belief that it was planted.

Potamogeton natans L. waters, sloughs and rivers V. 10 Atl.—Pacif.G.—
  60° N.L.
Potamogeton pauciflorus Ph. pools I. 5 N.E.—Ga.—N.Mex.—Up.Miss.
Potamogeton pusillus L. River I. 5 N.E.—R.Mts.—60° N.L.
Potamogeton pectinatus L. River V. 8 Atl.—Pacif.G.—55° N.L.

Alismaceae. In N.Am. 6 Gen. 15 Spec.

Triglochin palustre L. bottom springy places III. 8 N.Y.—Up.Miss. R.Mts.
  —Alaska—Greenland.
Triglochin maritimum L. var. elatum bottom springy places II, 4 N,Y.—Cal
  Labr.—Alaska.
Alisma Plantago L. shallow waters V. 5 N.E.—Ga.—Cal.—55° N.L.
Echinodorus rostratus Engelm. low banks III. 5 Fla.—Ariz.—Up.Miss.
Sagittaria variabilis Engelm. low banks, pools VII. 7 Atl.—Pacif.G.—Can.
  N.F.
Sagittaria calycina Engelm. swamps II. 3 N.E.—Up.Miss.
Sagittaria heterophylla Ph. swamps, low banks II. 8 N.E.—Fla.—Miss.

Hydrocharidaceae. In N.Am. 3 Gen. 3 Spec.

Anacharis canadensis Planch. creeks IV. 10 N.E.—N.Ca.—Up.Miss.—55°
  N.L.
Valisneria spiralis L. river IV. 10 Atl.—Miss.

Orchidaceae. In N.Am. 24 Gen. 99 Spec.

Orchis spectabilis L. wooded hillsides III. 4 N.E.—All. (Ga.)—Up.Miss.
Habenaria virescens Spr. wooded hillsides II. 3 Atl.—Miss.G.—Can.
†Habenaria hyperborea R.Br. bottom springy places I. 2 N.E.—Or. Alaska—
  Greenland.
Habenaria leucophæa Gr. wet prairie I. 5 Oh.—Up.Mo.
Spiranthes cernua Rich. moist banks II. 3 Atl.—W.Tex.—Or.G.—Can.
Spiranthes gracilis Big. dry hillsides III. 3 Atl.—Miss.G.—Can.
†Pogonia pendula Lindl. woods I. 3 Atl.—Miss.G.—Can.
Liparis Lœselii Rich. swamps I. 4 N.E.—Up.Miss.—54° N.L.
Liparis liliifolia Rich. shady hillsides I. 1 N.E.—Up.Miss.
Corallorhiza odontorhiza Nutt. woods I. 2 N.Y.—Fla.—R.Mts.
†Aplectrum hiemale Nutt. woods I. 3 N.E.—All.—Up.Mo.
Calopogon pulchellus R. Br. swamps I. 4 N.E.—Up.Miss. Fla.—La.
Cypripedium candidum Muhl. bottom II. 3 W.N.Y.—Up.Miss.
†Cypripedium parviflorum Salisb. bottom I. 2 N.Y.—All.—La.—Up.Miss.
Cypripedium pubescens Willd. woods V. 4 N.E.—All.—La.—R.Mts.
Cypripedium spectabile Sw. bottom, springy places II. 4 NE.—N.Ca.—Miss.

Amaryllidaceae. In N. Am. 5 Gen. 21 Spec.

Hypoxis erecta L. dry open woods, prairie V. 4 Atl.—R.Mts. G.—Can.

Iridaceae. In N. Am. 3 Gen. 21 Spec.

Iris versicolor L. bottom, banks V. 5 Atl.—Up.Mo. G.—Can.
Sisyrinchium Bermudiana L. prairie, open woods V. 6 Atl.—Pacif. G—
  Alaska.

Dioscoreaceae. In N. Am. 1 Gen. 1 Spec.

Dioscorea villosa L. woods, copses IV. 4 Atl.—Miss. G.—Can.

Smilacaceae. In N. Am. 1 Gen. 14 Spec.

Smilax hispida Muhl. moist copses, banks V. 5 W.N.Y.—La.—Up.Miss.
Smilax herbacea L. banks, bottom V. 4 Atl.—Up.Mo. G.—Can.

Liliaceæ. In N. Am. 50 Gen. 246 Spec.

Trillium recurvatum Beck. woods V. 5 Ind.—Up.Miss.

Trillium erectum L. var. album woods I. 2 N.E.—N.Ca.—Up.Miss.

Trillium nivale Ridd. woods IV. 4 Oh.—Up.Miss.

Uvularia grandiflora Sm. woods V. 4 Vt.—All. (Ga.)—Up.Mo.

Smilacina racemosa Desf. woods, copses V. 4 N.E.—All.—Cal. Or.—Sask.

Smilacina stellata Desf. bottom, banks III. 3 N.E.—N.Mex. Cal.—Arct.

Polygonatum giganteum Dietrich. woods V. 4 N.E.—La.—R.Mts.—Sask.

Lilium philadelphicum L. prairie, open woods V. 5 N.E.—N.Ca.—Up.Mo.

Lilium superbum L. copses, banks III. 3 N.E.—All. (Ga.)—La.—Up.Miss.

Erythronium albidum Nutt. woods V. 5 N.Y.—Up.Miss.

Scilla Fraseri Gr. prairie, banks V. 5 Oh.—La.—W.Tex.—Or.

Allium tricoccum Ait. shady hillsides III. 4 N.E.—N.Ca.—Up.Miss.

Allium canadense Kalm. wet prairie, banks V. 4 Atl.—Up Mo. G.—Can.

Juncaceæ. In N. Am. 2 Gen. 61 Spec.

Juncus tenuis Willd. bottom, woods, roadsides, etc. VII. 8 N.E.—Fla.—Cal.

Juncus acuminatus Michx. var. legitimus swamps V. 5 N.E.—Ga.—Ark.—
Up.Miss.

Juncus canadensis Gay. var. brachycephalus moist places V. 6 N.E.—
Up.Miss.—N.F.—Huds.

Juncus nodosus L. var. megacephalus swamps, banks IV. 4 N.F.—La.—
Cal.

Pontederiaceæ. In N. Am. 3 Gen. 4 Spec.

Pontederia cordata L. shallow waters, banks II. 10 Atl.—Miss. G.—Can.

Schollera graminea Willd. river, low banks VI. 8 N.E.—N.Ca.—Mex.

Commelynaceæ. In N. Am. 3 Gen. 12 Spec.

Tradescantia virginica L. prairie, copses V. 5 Atl.—W.Tex. G.—Can.

Commelyna cayennensis* Rich. banks I. 3 Up.Miss.—La.

Cyperaceæ. In N. Am. 20 Gen. 448 Spec.

Cyperus diandrus Torr. bottom V. 10 N.E.—N.Ca.—W.Tex.—Up.Miss.

Cyperus erythrorhizus Muhl. bottom V. 5 Pa.—Fla.—Ariz.—Up.Miss.

Cyperus inflexus Muhl. sandy banks V. 8 Atl.—N.Mex.—Cal. 52°N.L.

Cyperus acuminatus Torr. bottom III. 3 Ill.—Up.Mo.

Cyperus phymatodes Muhl. bottom, sandy banks V. 8 Vt.—Fla.—Ariz.—
Cal.

Cyperus strigosus L. bottom V. 6 Atl.—N.Mex. G.—Can.

Cyperus Michauxianus Schult. bottom V. 6 Atl.—Pacif. G.—Can.

Cyperus filiculmis Vahl. dry prairie, hillsides V. 4 Atl.—W.Tex. G.—Can.

†Cyperus ovularis Torr. sandy soil I. 2 N.Y.—Up.Miss. Fla.—W.Tex.

Dulichium spathaceum Pers. bottom, wet banks II. 8 Atl.—Miss.—Nebr.

Hemicarpha subsquarrosa** Nees. sandy banks V. 6 N.Y.—Fla. N. Mex.—
R.Mts.

Eleocharis obtusa Schult. bottom V. 8 N.E.—Up.Miss. Fla.—W.Tex.

Eleocharis palustris R. Br. swamps X. 10 Atl.—Pacif. G.—60°N.L.—Greenland.

Eleocharis compressa Sull. bottom III. 6 N.Y.—Up.Mo.

Eleocharis Wolffii Gr. bottom II. 6 Ill.

Eleocharis intermedia Schult. swamps, sandy banks V. 5 N.Y.—Ga.—Miss.

*Commelyna cayennensis is found on a single place in a wood. It is not uncommon in culti-
vation, and therefore it is doubtfull whether it is indigenous or not. It is common in South Illinois
and certainly an annual, although it may occasionally become perennial by striking roots from the
joints.
** Boeckeler (Linnaea 36 499) reunited this genus with Scirpus, and our species as Scirpus
micranthus Vahl., the oldest name.

Eleocharis tenuis Schult. swamps II. 4 N.E.—N.Ca.—Up.Miss.—W.Tex.
Eleocharis acicularis R. Br. low banks, shallow water VI. 10 Atl.—Pacif. G.
—55°N.L.
Scirpus pungens Vahl. banks V. 6 Atl.—N.Mex.—Cal. G.—Can.
Scirpus validus Vahl. banks V. 6 Atl.—N.Mex.—Cal. G.—Can.
Scirpus Smithii Gr. low banks, bottom II. 5 L. Ontario—Del.—Ill.
Scirpus atrovirens Muhl. bottom V. 4 N.E.—Ky —R.Mts.
Scirpus lineatus Michx. bottom II. 4 N.E.—Up.Mo. Fla.—Tex.
†Eriophorum gracile Koch. swamps I. 5 N.E.—Up.Mo. Fla.—Cal.—Arct.
Fimbristylis autumnalis R. Sch. gravelly banks III. 6 N.E.—Up.Mo. Fla.—
W.Tex.
Rhynchospora alba Vahl. swamps II. 5 Atl.—Up.Mo.—Alaska G.—60°N.L.
Scleria triglomerata Michx. wet prairie I. 3 Vt.—Fla.—Miss.
Scleria verticillata* Muhl. swamps I. 4 W.N.Y.—Fla.—Mich.—Up.Miss.
Carex polytrichoides Muhl. bottom VI. 10 Atl.—R.Mts. G.—Sask.
Carex Steudelii Kth. woods IV. 8 N.Y.—Fla.—Miss.
Carex disticha Huds. sandy prairie III. 3 N.Y.—R.Mts.—Cal. Ill.—Sask.
Carex teretiuscula Good. moist soil V. 5 N.E.—Or.—53°N.L.
Carex vulpinoidea Michx. bottom VI. 6 N.E.—S.Ca.—Up.Mo.
Carex crus corvi Shuttl. wet prairie III. 3 Oh.—Wisc.—La.—W.Fla.
Carex stipata Muhl. wet prairie III. 5 N.E.—Fla.—La.—Or.--54°N.L.
Carex conjuncta Boot. bottom V. 5 N.E.—All. (Ga.)--Up.Miss.
Carex sparganioides Muhl. wooded hillsides VI. 6 N.E.—All. (Ga.)—Up.
Miss.
Carex cephaloidea Boot. woods II. 4 N.Y.—Ill.
Carex cephalophora Muhl. open woods V. 5 N.E.—Up.Miss. Fla.—W.Tex.
Carex rosea Schk. woods VI. 6 N.E.—All. (Ga.)—Up.Mo.—Or.
Carex sterilis Willd. bottom II. 5 N.E.—Fla.—Up.Miss.
Carex stellulata Good. woods IV. 5 Atl.—Or.—Alaska G.—54° N.L.
Carex arida Schw. Torr. bottom II. 6 Ky.—Up.Miss.—54° N.L.
Carex scoparia Schk. bottom V. 6 N.E.—N.Ca.—Or.—54° N.L.
Carex lagopodioides Schk. bottom V. 5 N.E.—S.Ca.—Cal.—54° N.L.
Carex cristata Schw. bottom V. 7 Del. Pa.—Up.Mo.—54° N.L.
Carex straminea Schk. bottom VI. 7 N.E.—Or. Fla.—W.Tex.
Carex stricta Lam. swamp III. 7 N.E.—All. (N.Ca.)—N.Mex.—Arct.
Carex limosa L. swamp I. 3 N.E.—R.Mts. Or.—G. Bear Lake.
Carex Buxbaumii Wahl. dry prairie I. 3 N.E.—All. (Ga.)—Tex.—Cal.—Alaska.—Huds.
Carex Shortiana Dew. banks, shady woods V. 5 Pa.—Va.—Up.Mo.
Carex tetanica Schk. var. Meadii Dew. dry prairie III. 5 Oh. Ill. Wisc.
Carex granularis Muhl. wooded hillsides VI. 6 Atl.—Miss.G.—Can.
Carex grisea Wahl. wooded hillsides VI. 6 Atl.—Up.Mo.—Tex.G.—Can.
Carex Davisii Schw. Torr. woods V. 5 N.E.—All. (Ga.)—Up.Mo.
Carex triceps Michx. open woods III. 5 N.E.—Fla.—Miss.—Tex.
Carex digitalis Willd. wooded hillsides I. 3 N.Y.—Ky.—Up.Miss.
Carex laxiflora Lam. woods VI. 6 Atl.—Up.Mo.—Or.G.—54° N.L.
Carex oligocarpa Schk. woods II. 3 N.E.—Ky.—Up.Miss.
†Carex Hitchcockiana Dew. woods I. 2 N.E.—Ky.—Up.Miss.
Carex umbellata Schk. dry hillsides I. 4 N.E.—Ill.—R.Mts.—Sask. La.—Ariz.
Carex pennsylvanica Lam. woods VIII. 8 N.E.—All. (Ga.)—R.Mts.
Carex varia Muhl. woods V. 5 N.E.--Up.Miss.
†Carex Richardsonii R.Br. open woods I. 2 N.Y.--Up.Miss. 54° N.L.—N.W.
Coast.
Carex pubescens Muhl. moist woods V. 5 N.E.--Up.Miss.

---

* This rare plant was found during the summer of 1887 by Mr. McDonald, the first time in Illinois.

Carex filiformis L. swamps III. 5 N.E.—Up.Mo.—54° N.L.
Carex lanuginosa Michx. wet prairie V. 5 N.E.—Ky.—N.Mex.—Cal.—Mc-
Kenzie.
Carex riparia Curt. swamp I. 5 Fla. N.E.—Up.Miss.
Carex trichocarpa Muhl. bottom II. 5 N.E.—Ga.—Up.Miss.
Carex comosa Boot. swamp I. 4 N.E.—Up.Miss.
Carex histricina Willd. wet banks, bottom VII. 7 N.E.—Up.Miss. Ga.—
N.Mex.
Carex tentaculata Muhl. bottom VII. 6 Atl.—Miss.G.—Can.
Carex Grayi Carey bottom V. 4 N.Y.—Up.Miss.
Carex lupulina Muhl. bottom V. 5 Atl.—Miss.G.—Can.
Carex lupuliformis Sartwell bottom I. 3 N.Y.—Del.—Ill.
Carex monile Tuck. swamps II. 3 N.E.—Ky.—65° N.L.
Carex squarrosa L. bottom I. 4 N.E.—All. (Ga.)—Up.Miss.
Carex longirostris Torr. woods III. 5 N.E.—R.Mts.—54° N.L.

Gramineæ. In N. Am. 103 Gen. 615 Spec.

Leersia oryzoides Sw. banks V. 7 Atl.—Miss. G.—54° N.L.
Leersia virginica Willd. bottom V. 5 Atl.—Miss. G.—Can.
Leersia lenticularis Michx. bottom II. 4 Va.—Fl.—La.—Up.Miss.
Zizania aquatica L. swamps, shallow waters IV. 8 Atl.—Miss. G.—Can.
Alopecurus geniculatus var. aristulatus Michx. fields, waste places III. 3
N.E.—Fla.—Up.Miss.
Vilfa aspera Beauv. sandy soil IV. 4 Atl.—Miss.
Vilfa vaginaeflora Torr. sandy soil VI. 5 N.E.—N.Ca.—Ark.—Up.Miss.
Sporobolus heterolepis Gr. dry hillsides V. 6 N.E.—Up.Miss.
Agrostis perennans Tuck. bottom IV. 5 Atl.—Miss.
Agrostis scabra Willd. dry prairies III. 5 N.E.—Cal.—Alaska Pa.—La.
Agrostis vulgaris With. open woods V. 5 N.Y.—Ark.—R.Mts.—55° N.L.
Agrostis alba*L. bottom IV. 3 N.Y.—Cal.—Huds.
Cinna arundinacea L. bottom, banks V. 5 Atl.—Miss. G.—55° N.L.
Muhlenbergia sobolifera Gr. woods II. 3 Vt.—Up.Miss.
†Muhlenbergia glomerata Trin. swamps I. 4 N.E.—Up.-Miss.—Ark.—Sask.
—Or.
Muhlenbergia mexicana Trin. bottom V. 6 N.E.—N.Ca.—Ark.—Up.Miss.
Muhlenbergia silvatica T. Gr. bottom III. 6 N.E.—Up.Miss.—W.Tex.—
Nevada.
Muhlenbergia Willdenovii Trin. woods II. 3 N.E.—All. (Ga.)—Up.Miss.
Muhlenbergia diffusa Schrad. bottom VI. 8 Atl.—Cal. G.—Can.
Brachyelytrum aristatum Beauv. bottom III. 4 Atl.—Miss.
Calamagrostis canadensis Beauv. bottom II. 7 N.E.—All. (Ga.)—R.Mts.—
Arct.
Oryzopsis melanocarpa Muhl. sandy hillsides I. 4 N.E.—Up.Miss.
Stipa spartea Trin. prairie V. 5 Ill.—N.Mich.—Up.Mo.—R.Mts.—Nevada.
Spartina cynosuroides Willd. bottom V. 10 N.E.—Up.Mo.—Arct.
Bouteloua curtipendula Gr. dry prairie VI. 6. N.Y.—Mex.
Tricuspis seslerioides Torr. dry prairie IV. 5 N.Y.—Fla.—La.—Up.Miss.
Diarrhena americana Beauv. shady woods IV. 4 Oh.—Up.Miss.—Ark.
Koeleria cristata Torr. dry prairie V. 7 Pa.—Cal.—Or.—54° N.L.
Eatonia obtusata Gr. dry prairie III. 5 Pa.—Fla.—Cal.—Or.
Eatonia pennsylvanica Gr. woods V. 5. N.E.—All. (Ga.)—Up.Mo.
Melica mutica Walt. copses, woods V. 4 W.Pa.—Fla.—R.Mts.
Glyzeria nervata Trin. banks VI. 8 Atl.—Miss.
Glyzeria fluitans R.Br. creeks III. 6 N.E.—All. (Ga.)—Miss.—54 N.L.

Poa sylvestris Gr. woods III. 5 Mich.—Ky.—Up.Miss.
Poa serotina Ehrh. banks II. 4 N.E.—R.Mts.—Or.—Alaska.
Poa pratensis L. prairies, cultivated ground V. 10 N.E.—Up.Mo.—72° N.L.
    Kotzebue Sund—Greenland.
Poa compressa L. dry soil V. 8 N.E.—Up.Miss.—54° N.L.
Poa annua L. dry prairie, cultivated land V. 4 Atl.—Pacif.
Eragrostis reptans Nees low sandy banks V. 10 Atl.—Up.Miss.
Eragrostis Frankii Mey. sandy banks V. 8 Oh.—Up.Miss.
Eragrostis capillaris Nees sandy soil, fields V. 8 Atl.—Miss.
Eragrostis pectinacea Gr. var. spectabilis sandy soil V. 7. Mass.—Oh.—Fla,
    —Miss.
Festuca tenella Willd. dry prairie V. 6. N.E —Up.Mo. Fla.—Cal.
Festuca nutans Willd. woods V. 5 N.E.—Fla.—Up.Mo.
Bromus Kalmii Gr. woods IV. 4 N.E.—Cal.—Arct.
Bromus ciliatus L. woods IV. 4 Atl.—Pacif. G.—Arct.
Phragmites communis Trin. banks, swamps II. 6 Atl.—Pacif. G.—54°N.L.
Hordeum pratense Huds. roadsides IV. 5 Oh.—La.—R.Mts.
*Hordeum jubatum L. fields I. 3 W.Can.—Up.Miss.—R.Mts.—Cal.
Elymus virginicus L. woods IV. 4 Atl.—Up.Mo. G.—Can.
Elymus canadensis L. prairie, copses III. 5 N.E.—Up.Mo.
Elymus strictus Willd. var. villosus woods V. 4 N.E.—All.—Cal.
Gymnostichum Hystrix Schreb. woods V. 3 N.E.—All. (Ga.)—Up.Miss.
Danthonia spicata Beauv. dry open woods III. 6 Atl.—Miss.
Phalaris arundinacea L. swamps II. 6 N.E.—Up.Miss.—Cal.—60°N.L.
Panicum anceps Michx. moist places V. 5 N.J.—Up.Miss. Fla.—La.
Panicum proliferum Lam. roadsides, moist places V. 6 Atl.—Miss.
Panicum capillare L. sandy soil V. 7 Atl.—Cal. G.—Can.
Panicum autumnale Bosc. sandy prairie I. 5 S.Ca.—Ill.
Panicum virgatum L. banks, bottom V. 4 N.E.—Up.Mo. Fla.—N.Mex.
Panicum clandestinum L. copses II. 4 N.E.—N.Ca.—Up.Mo.
Panicum latifolium L. copses V. 4 N.E.—Up.Miss. Fla.—N.Mex.
Panicum scoparium Lam. prairie III. 4 N.W.—Ga.—Miss.
Panicum dichotomum L. open woods, prairie V. 7 N.E.—Cal. Fla.—W.Tex.
Panicum depauperatum Muhl. dry prairie, copses III. 5 N.E.—N.Ca.—
    Up.Miss.
Panicum crus Galli L. bottom, fields VII. 7 N.E.—Fla.—Tex.—Cal.—Or.
Cenchrus tribuloides L. bottom, cultivated land V. 5 N.E.—Fla.—Tex.—
    Cal.—Up.Mo.
Andropogon furcatus Muhl. prairie V. 5 N.E.—Up.Mo. Fla.—N.Mex.
Andropogon scoparius Michx. sandy soil V. 7 N.E.—Up.Mo. Fla.—N.Mex.
Chrysopogon nutans Benth. dry prairie V. 6 N.E.—Up.Mo. Fla.—N.Mex.

Equisetaceæ. In N. Am. 1 Gen. 13 Spec.
    Equisetum arvense L. sandy soil, banks VI. 3 N.E.—N.Mex.—Cal.—Arct.
    †Equisetum palustre L bottom I. 5 N.Y.—Up.Miss.
    Equisetum limosum L. shallow waters VI. 4 N.E.—Up.Miss.
    Equisetum laevigatum A. Br. dry clay soil I. 3 N.Ca.—Ill.—Cal.
    Equisetum hiemale L. banks VI. 5 N.E.—N.Mex.—Cal.
    Equisetum robustum A. Br. banks V. 6 Oh.—N.Mex.—Or.
    Equisetum variegatum Schleich. banks II 3 N.E.—R.Mts.—72°N.L.

Filices. In N. Am. 30 Gen. 146 Spec.
    Adiantum pedatum L. woods VIII. 8 N.E.—N.Ca.—Cal.—Alaska.

---

*Hordeum jubatum is observed only since five years. When not overlooked before, as it is not common, then it may be that it recently immigrated from the Northwest.

Pteris aquilina L. woods V. 5 N.E.—Alaska Fla.—Ariz.
Asplenium angustifolium Michx. woods II. 3 N.E.—All. (Ga.)—Ark.-
Up.Mo.
Asplenium filixfoemina Bernh. woods V.5 Labr. N.E.—Fla.—Cal.—Alaska.
Asplenium thelypteroides Michx. woods III. 5 N.E.—All. (Ga.)—La.—
Up.Miss.
Camptosorus rhizophyllus Link. rocks II. 3 N.E.—All. (Ga.)—Wisc.
Phegopteris hexagonoptera Feé. woods III. 5 Atl.—Miss.
Aspidium thelypteris Sw. swamps V. 10 Atl.—Miss. G.—52°N.L.
Aspidium spinulosum Sw. woods III, 4 N.F.—N.Ca.—Ark.—Or.—Alaska—
Arct.
Aspidium acrostichoides Sw. woods V. 5 Atl.—Miss.
Cystopteris bulbifera Bernh. shaded rocks III. 5 N.E.—N.Ca.—Ark.- Up.
Miss.
Cystopteris fragilis Bernh. woods VII. 8 N.E.—N.Ca.—Cal—Alaska—Green-
land 72°N.L.
Onoclea sensibilis L. woods III. 5 N.Br.—Fla. G.—Sask.
†Struthiopteris germanica Willd. woods I. 1 N.E.— All. (Ga.)—Sask.
†Osmunda regalis L. swampy bottom I. 2 Atl.—Miss. N.F.—Sask.
Osmunda claytoniana L. woods V. 6 N.F.—All.(Ga.)—Ark.—Lake Superior.

Ophioglossaceæ. In N. Am. 2 Gen. 11 Spec.
Botrychium virginicum Sw. woods IV. 4 Atl.—Or.

Lycopodiaceæ. In N. Am. 3 Gen. 20 Spec.
Selaginella apus Spring. moist places near springs III. 5 Atl.—Miss.

Hydropterides. In N. Am. 4 Gen. 22 Spec.
Azolla caroliniana Willd. river, wet banks I. 10 Atl,—Miss.—Ariz.

Musci. In N Am. 120 Gen. 896 Spec.
Phascum cuspidatum Schreb., old fields.
Pleuridium alternifolium Brid., old fields.
Weisia viridula Brid., old fields, roadsides.
Dicranella varia Schpr., clay soil.
Dicranella heteromalla Schpr., moist ground.
Dicranum scoparium L., wooded hillsides.
Fissidens bryoides Hdw., moist ground in woods.
Fissidens taxifolius Hdw., on sandy soil in woods.
Fissidens adiantoides Hdw., moist ground in woods.
*Fissidens subbasilaris Hdw., on putrid trunks.
Leucobryum vulgare Hampe, moist ground.
Ceratodon purpureus Brid., on the ground in woods and fields.
Didymodon rubellus Br. Sch., on sandy ground.

*The species marked (*) are not common with the Eastern Continent. Fissidens grandifrous
Brid. and Gymnostomum curvirostrum Hedw., before not known in Illinois, I found in the canyon
called Deer Park, on the Vermillion river, in LaSalle county in 1883.
A list of Illinois mosses, published in 1878 by Hall and Wolf in Bulletin II. of Illinois State
Laboratory, contains 50 genera and 153 species. No doubt there will be found more, although the
nature of the country, the little difference in elevation and soil, and the lack of coniferous forests is
not favorable to moss growth.
A great number found by Wolf in Fulton county have not yet been observed around Peoria,
but a greater part may occur. These are the following: Ephemerum crassinervium Hampe, Sphaer-
angium Schimperianum Lesq., Astomum nitidulum Schimp., Bruchia flexuosa C. Muell., Weisia
Wolfii James, Campylopus Leanus Sull., Dicranella rufescens Schimp., Dicranum flagellare Hedw.,
Fissidens obtusifolius Wils., Fissideus minutulus Sull., Conomitrium Julianum Mort., Ulota crispula
Brid., Atrichum undulatum Beauv., Webera nutans Hedw., Webera albicaus Schimp., Bryum pendu-
lum Schimp., Bryum atropurpureum Wahl., Mnium affine Blaud., Philonotis Mühlenbergii Brid.,
Discelium nudum Brid., Aphanorhegma serrata Sull., Fontinalis biformis Sull., Fontinalis disticha
Hook., Fontinalis dalecarlica Br. Eu., Dichelyma capillaceum Br. Eu., Leskea Austini Sull., Clasma-
todon parvulus Hampe, Thelia Lescurii Sull., Fabronia gymnostoma Sull. and Lesq., Anacamptodon
splachnoides Brid., Pylaisia denticulata Schimp., Cylindrothecium compressum Br. Eu., Hypnum
triquetrum L., Hypnum Sullivantii Spruce., Hypnum Boseii Schwaegr., Hypnum micans var. albu-
lum C. Muell.. Hypnum crista castrensis L.. Hypnum dimorphum Brid.. Hypnum acutum Mitt.

Leptotrichum pallidum Hampe, clayey ground.
Leptotrichum tortile C. Muell., clayey ground.
Desmatodon obtusifolius Schpr., on sandy soil.
Barbula unguiculata Hdw., clayey ground.
Barbula caespitosa Schwaegr., on exposed roots of trees.
Grimmia conferta Funk., on rocks.
Hedwigia ciliata Ehrh., on rocks.
*Drummondia clavellata Hook., on bark of trees.
*Orthotrichum strangulatum Beauv., on bark of trees.
Physcomitrium pyriforme Brid., on the ground.
Funaria hygrometrica Sibth., on the ground.
Bartramia pomiformis Hedw., on the ground in shady woods.
Leptobryum pyriforme Schimp., on the ground.
Bryum intermedium Brid., on clayey soil.
Bryum bimum Schreb., on the ground in woods.
Bryum argenteum L., on the ground around houses.
Bryum caespitosum L., on the ground.
Bryum roseum Schreb., in shady places.
Bryum uliginosum Br. Sch., in woods.
Mnium cuspidatum Hedw., in woods around trees.
*Aulacomnium heterostichum Br. Sch., in the shade of trees.
Timmia megapolitana Hedw., on sandy soil.
Atrichum angustatum Br. Sch., on the ground in woods.
*Pogonatum brevicaule Beauv., on clayey soil.
Polytrichum juniperinum Willd., on the ground in woods.
Polytrichum commune L., on the ground in woods.
*Leucodon julaceus Sull., at the base of trees.
*Thelia hirtella Sull., at the base of trees.
*Thelia asprella Sull., at the base of trees.
Leskea polycarpa Ehrh., at the base of trees.
Leskea obscura Hedw., at the base of trees.
Anomodon rostratus Schimp., on the ground in woods.
Anomodon obtusifolius Br. Sch., on trunks of trees.
Anomodon attenuatus Hueb., on trunks of trees.
Anomodon tristis Cesati., on trunks of trees.
Platygyrium repens Br. Sch., on the bark of decayed logs.
Pylaisia intricata Br. Sch., on the bark of trees.
Pylaisia velutina Br. Sch., on the bark of trees.
Homalothecium subcapillatum Br. Eu., on the bark of trees.
Cylindrothecium cladorrhizans Schimp., on old logs.
*Cylindrothecium sedutrix Sull., on old logs.
*Climacium americanum Brid., on the ground in woods.
Hypnum minutulum Hedw., on decayed logs.
*Hypnum gracile Br. Sch., on decayed logs.
Hypnum delicatulum L., on decayed logs.
Hypnum laetum Brid., on exposed roots of trees.
*Hypnum acuminatum Beauv., on the ground in woods.
Hypnum salebrosum Hoffm., on the ground in woods.
Hypnum rutabulum L., on the ground in woods.
Hypnum rivulare Br. Eu., on moist ground in woods.
Hypnum hians Hedw., on the ground in woods.
*Hypnum serrulatum Hedw., on the ground in woods.
*Hypnum deplanatum Schpr., on the ground in woods.
*Hypnum adnatum Hedw., on boulders and trunks.

Hypnum serpens L., on the ground and rotten logs.
Hypnum radicale Beauv., on the ground and rotten logs.
Hypnum orthocladum Beauv., around springs.
Hypnum riparium L., in swampy places.
*Hypnum hispidulum Brid., in dry woods on the ground or logs.
Hypnum chrysophyllum Brid., in woods on the ground.
Hypnum aduncum Hedw., in swamps.
Hypnum imponens Hedw., on decayed logs.
Hypnum curvifolium Hedw., on decayed logs.
Hypnum Haldanianum Grev., in woods on the ground.
Hypnum Schreberi Willd., in damp woods on the ground.

Hepaticæ. In N.America, 50 genera, 230 species.

The list of Illinois Hepaticæ by Wolf and Hall contains 44 species in 22 genera. There are seven species that are mentioned of Fulton County and may occur around Peoria, although not yet observed, viz: Riccia Lescuriana Aust., Aneura pinguis Dum., Aneura multifida Dum., Madotheca porella Nees., Lophocolea Macouni Aust., Coleochila Taylori Dum., Blepharostoma trichophylla Dum.

*Riccia Frostii Aust., on moist ground.
*Riccia lutescens Schwein., on moist ground.
Riccia fluitans L., in waters and moist ground.
Riccia natans L., in water.
Marchantia polymorpha L., on moist ground in woods.
Grimaldia barbifrons Bisch., in woods between mosses.
Asterella hemisphaerica Beauv., on moist ground.
Conocephalus conicus Dum., on wet rocks, moist hillsides.
*Fimbriaria tenella Nees., on the ground in woods.
Anthoceros laevis L., on wet boulders between mosses.
*Notothylas orbicularis Sull., on moist ground along creeks.
*Frullania eboracensis Gottsche., on the bark of trees.
*Frullania virginica Gottsche., on the bark of trees.
*Frullania aeolotis Nees., on decayed logs.
Madotheca platyphylla Dum., on the bark of trees.
Radula complanata Dum., on the bark of trees.
Blepharozia ciliaris Dum., on old logs.
Trichocolea tomentella Dum., in moist places in woods.
Lophocolea bidentata Dum., between mosses in woods.
Lophocolea heterophylla Nees., between mosses in woods.
Cephalozia bicuspidata Dum., on rotten logs.
Cephalozia divaricata Dum., on rotten logs.
Cephalozia curvifolia Dum., on rotten logs.
Jungermannia Schraderi Mart., on rotten logs.

The Lichens, Fungi and Algae of our flora are little studied; though the few species identified may be mentioned.

Lichenes*.

Usnea barbata Fr.
Ramalina calicaris Fr.
Evernia jubata Fr.
Theloschistes parietina Norm.
Parmelia perforata Ach.
Parmelia perlata Ach.
Parmelia olivacea Ach.
Physcia pulverulenta Schreb.
Physcia speciosa Wulfen.
Physcia stellaris L.
Physcia caesia Hoffm.
Physcia obscura Ehrh.
Peltigera canina Hoffm.
Lecanora varia Fr.
Lecanora subfusca Ach.
Placodium aurantiacum Lightf.
Placodium ferrugineum Huds.
Placodium cerinum Ach.

Pertusaria velata Turn.
Calicium subtile Fr.
Cladonia pyxidata Fr.
Cladonia fimbriata Fr.
Cladonia gracilis Fr.
Cladonia mitrula Tuck.
Cladonia furcata Floerke.
Cladonia macilenta Hoffm.
Cladonia rangiferina Hoffm.
Biatora campestris Fr.
Biatora flexuosa Fr.
Biatora rubella Fr.
Biatora sanguineo-atra Fr.
Opegrapha varia Pers.
Graphis scripta Ach.
Collema nigrescens Huds.
Leptogium pulchellum Ach.

Fungi.

Agaricus procerus Scop.
Agaricus decolorans Mich.
Agaricus radicatus Relhan.
Agaricus velutipes Curt.
Agaricus pyxidatus Bull.
Agaricus domesticus Bolt.
Agaricus cinereus Bull.
Agaricus campestris L.
Agaricus rosaceus Nees.
Agaricus laccatus Scop.
Agaricus ostreatus Jacq.
Agaricus campanella Batsch.
Agaricus flabelliformis Bolt.
Lenzites tigrinus Fr.
Lenzites vialis Peck.
Marasmius Rotula Scop.
Schizophyllum commune Fr.
Favolus canadensis Klotsch.
Polyporus lucidus Fr.
Polyporus badius Schw.
Polyporus versicolor L.
Polyporus cinnabarinus Jacq.
Polyporus picipes Fr.
Polyporus gilvus Schw.
Polyporus fomentarius L.

Polyporus igniarius L.
Polyporus arcularius Fr.
Polyporus varius Fr.
Trametes Peckii.
Merulius lacrimans Fr.
Merulius spathularia Schw.
Clavaria aurea Schaeff.
Clavaria botrytis Pers.
Clavaria cristata Holmsk.
Clavaria fastigiata D.C.
Tremella foliacea Pers.
Tremella lutescens Fr.
Helvella esculenta Pers.
Helvella crispa Fr.
Morchella esculenta Pers.
Phallus Ravenelii B.&C.
Geaster mammosus Fr.
Stemonitis fusca Roth.
Hemiarcyria rubiformis Pers.
Cyathus vernicosus D.C.
Crucibulum vulgare Tul.
Peziza badia Pers.
Peziza aurantia Vahl. .
Peziza coccinea Jacq.
Peziza umbrina Pers.

*The first part of Tuckerman's synopsis of N. Amer. Lichens containing Parmeliacei, Cladoniei and Coenogoniei enumerates 46 Genera 411 species and 58 subspecies, of which after Wolf and Hall 22 genera with 93 species are found in Illinois. Of the rest of the genera the first edition of Tuckerman's Lichens (1848) contains only 16 genera and 109 species, and of those Wolf and Hall mention as occurring in Illinois 75 species, so that 168 species so far are known from Illinois.

Peziza acetabulum L.
Peziza luculenta Cke.
Bovista nigrescens Pers.
Lycoperdon pyxiforme Schaeff.
Uromyces Lespedezae Schw.
Uromyces Hedysari paniculati Farl.
Uromyces Polygoni Fckl.
Uromyces Howei Peck.
Uromyces Caladii Farl.
Uromyces Euphorbiae C.&P.
Uromyces graminicola Andr.
Ustilago Maydis Cd.
Gymnosporangium macropus L.N.
Phragmidium fragariae Rostk.
Phragmidium mucronatum Pers.
Coleosporium Sonchi-arvensis Lev.
Melampsora salicina Lev.
Caeoma agrimoniae Schw.
Caeoma nitens Schw.
Puccinia Podophylli Schw.
Puccinia violae D.C.
Puccinia Amorphae Curt.
Puccinia Circaeae Pers.
Puccinia galiorum Lk.
Puccinia Kuhniae Schw.
Puccinia Asteris Duby.
Puccinia Silphii Schw.
Puccinia Tanaceti D.C.
Puccinia flosculorum Roehl.
Puccinia Lobeliae Gerard.

Puccinia lateripes B.&C.
Puccinia Menthae Pers.
Puccinia aculeata Schw.
Puccinia Polygoni amphibii Pers.
Puccinia Smilacis Schw.
Puccinia caricis Rebent.
Puccinia graminis Pers.
Puccinia Andropogi Schw.
Puccinia Prunorum Lk.
Puccinia phragmites Kornike.
Roestelia pennicillata Fr.
Roestelia lacerata Fr.
Aecidium Ranunculi Schw.
Aecidium Pteleae B.&C.
Aecidium Psoraleae Peck.
Aecidium grossulariae D.C.
Aecidium oenotherae Peck.
Aecidium compositarum Burr.
Aecidium Euphorbiae Gm.
Aecidium Erigeronatum Schw.
Aecidium Podophylli Schw.
Cystopus cubicus Lev.
Cystopus candidus Lev.
Cystopus Bliti Lev.
Microsphaera Frisii Lev.
Entyloma Menispermi Farl.
Peronospora Arthuri Farl.
Peronospora viticola De.Bary.
Peronospora Halstedii Farl.

Characeæ.

Chara gymnopus A.Br.
Chara contraria A.Br.

Nitella acuminata var subglomerata A.Br.

*Algæ.

Oscillaria nigra Vauch.
Nostoc pruniforme Agh.
Palmella hyalina Lyngby?
Hydrodictyon utriculatum Roth.
Spirogyra quinina Kuetz.
Spirogyra setiformis Kuetz.

Hydrogastrum granulatum Desv.
Vaucheria spec.
Cladophora glomerata Rabenh.
Draparnaldia plumosa Agh.
Chaetophora elegans Agh.
Batrachospermum moniliforme Roth.

*Wood in "Fresh water Algæ of North America" 1872 enumerates 46 genera, 411 species and 57 subspecies.

COMPARATIVE STATISTICS.

The flora of Peoria contains 835 indigenous species of vascular plants, of which 139 are monocarpic, 583 rhizocarpic, and 113 woody. Possibly 30 or 40 species more may be found — most of them in water and swamps. This for so small an area, with a soil which is prevailingly arenaceous and humus, only partly argillaceous, with very little calcareous matter in it, is not a poor flora compared with those of larger areas. The flora of Illinois *, when we exclude 29 species restricted to the shores of Lake Michigan, which form a part of the Canadian flora, and 56 species of the most southern part of the state along the Ohio and Wabash, belonging to the flora of the Ohio Valley, numbers 1,353 species of 525 genera in 122 orders; 504 species have not yet been found in the vicinity of Peoria, 220 of which are reported only from the southward, 131 from northward, 7 only from westward, and 158 as well north- as southward.

*The species excluded from the Illinois flora proper, so far only observed on the shore of Lake Michigan and along the Ohio and Wabash, are the following: On the shore of Lake Michigan — Ranunculus cymbalaria Ph., Sarracenia purpurea L., Cakile americana Nutt,, Hypericum kalmianum L., Drosera rotundifolia L., Drosera longifolia L., Lathyrus maritimus Big., Potentilla anserina L., Linnaea borealis Gron., Aster macrophyllus L., Solidago virga-aurea L. var. humilis, Solidago Muhlenbergii T. Gr., Silphium trifoliatum L., Cirsium Pitcheri T. Gr., Corispermum hyssopifolium L., Rumex maritimus L., Euphorbia polygonifolia L., Salix adenophylla Hook., Juniperus sabina L. var. procumbens. Potamogeton praelongus Wulf., Habenaria Hookeri Torr., Habenaria lacera R. Br., Medeola virginica L., Juncus balticus Deth., Juncus Gerardi Lois., Eriophorum virginicum L., Carex aurea Nutt., Carex Crawei Dew., Calamagrostis arenaria Roth.; along the Ohio and Wabash — Magnolia acuminata L., Cocculus carolinensis D. C., Calycocarpum Lyoni Nutt., Nuphar sagittifolia Ph., Cabomba caroliniana Gr., Corydalis flavula Raf., Draba brachycarpa Nutt., Elodes petiolata Ph., Cassia obtusifolia L. Myriophyllum ambiguum Nutt., Jussiaea repens L., Jussiaea decurrens D. C., Ludwigia cylindrica Ell., Galium lanceolatum Torr., Mikania scandens L. Solidago odorata Ait., Iva ciliata Willd., Verbesina occidentalis Walt., Anaphalis margaritacea Benth. Hook., Rhododendron nudiflorum Torr., Hottonia inflata Ell., Catalpa speciosa Ward., Epiphegus virginiana Bart , Verbena Aubletia L., Scutellaria serrata Andr., Pycnanthemum muticum Pers. (forma typica), Hydrophyllum macrophyllum Nutt., Phacelia Purshii Buckl., Hydrolea affinis Gr., Obolaria virginica L., Trachelospermum difforme Gr., Gonolobus laevis Michx., Iresine celosioides L., Brunnichia cirrhosa Banks, Acalypha caroliniana Walt., Tragia macrocarpa Willd., Quercus lyrata Walt., Quercus Phellos L., Fagus ferruginea Ait., Taxodium distichum Rich., Cupressus thyoides L., Najas indica var. gracillima A. Br., Goodyera repens R. Br., Iris cuprea Ph., Commelyna erecta L., Heteranthera reniformis R. P., Heteranthera limosa Vahl., Cyperus flavescens L., Fuirena squarrosa Michx., Rhynchospora corniculata Gr., Carex gigantea Rudge, Arundinaria tecta Muhl., Leptochloa mucronata Kth., Dicksonia punctilobula Kunze, Botrychium lunarioides Sw. var. obliquum, Ophioglossum vulgatum L.

A number of local floras and such of larger districts of the same geographical latitude or longitude are given in the following table, but it must be remarked that all mere varieties and all introduced species are excluded, and that the figures mean the number of the entire states, even if they, like New York state, Ohio, Illinois, and Wisconsin, belong to different natural floral districts:

	Massachusetts	New York.	Ohio.	Chester Co. Pa. Newcastle Co., Del.	Washington, D.C.
Area in Sq. Miles.	7,800	47.000	40,000	128	108
Vascular Plants:					
Genera.....	443	533	453	436	426
Species.....	1162	1330	1232	981	922

	Illinois.	Colorado.	Michigan. Lower Peninsula.	Wisconsin	Arkansas.	Louisiana.
Area in Square Miles...	55,400	104.500	33,400	53,900	52,200	41,300
Vascular Plants:						
Genera........	551	430	446	450	562	588
Species	1431	1145	1094	1104	1233	1555

The comparatively large number of species in Illinois is readily explained by the wide extension of this state in a south-north direction over 5° of latitude. The number of species in *Michigan, Wisconsin, Arkansas, and particularly in Colorado may really show greater figures, as those states are apparently not so thoroughly explored in all parts.

Generally the eight or ten largest orders make up half of the whole number of vascular species of a floral district. The following tables will show the per mille of species for each of the largest orders:

Massachusetts.		*New York.*		*Ohio.*	
Cyperaceæ..........	110	Cyperaceæ..........	113	Compositæ.........	122
Compositæ.........	108	Compositæ..........	104	Cyperaceæ	95
Gramineæ..........	73	Gramineæ..........	79	Gramineæ	65
Rosaceæ............	38	Rosaceæ............	36	Leguminosæ	40
Ericaceæ...........	33	Leguminosæ	34	Rosaceæ............	36
Filices	32	Ericaceæ...........	30	Orchidaceæ	32
Leguminosæ	26	Filices.............	30	Filices	30
Orchidaceæ.........	26	Orchidaceæ.........	29	Ranunculaceæ	29
Scrophulariaceæ	24	Scrophulariaceæ ...	25	Labiatæ............	28
Labiatæ	22	Labiatæ	25	Liliaceæ	28
	498		505		505

**Michigan means here the lower peninsula, and the flora is taken from the first Report of Geological Survey of Michigan, 1861. The flora of Wisconsin after Lapham's Catalogue in Transactions of the Wisconsin State Agricultural Society, 1853, with two additions by Hale. The flora of Arkansas is after Lesquereux' Catalogue in Arkansas Geological Survey (1860). The flora of Colorado is after Porter and Coulter (1874).

Chester, Pa.; New- castle, Del.	
Compositæ	130
Cyperaceæ	83
Gramineæ	78
Leguminosæ	40
Rosaceæ	33
Filices	32
Orchidaceæ	29
Labiatæ	28
Ericaceæ	26
Scrophulariaceæ	23
	502

Michigan.	
Compositæ	116
Cyperaceæ	100
Gramineæ	73
Rosaceæ	46
Leguminosæ	38
Orchidaceæ	35
Scrophulariaceæ	28
Ranunculaceæ	26
Filices	25
Labiatæ	25
	512

Wisconsin.	
Compositæ	123
Cyperaceæ	90
Gramineæ	72
Rosaceæ	40
Leguminosæ	38
Filices	30
Orchidaceæ	26
Scrophulariaceæ	26
Ranunculaceæ	26
Liliaceæ	24
	495

Arkansas.	
Compositæ	164
Gramineæ	78
Leguminosæ	77
Cyperaceæ	46
Labiatæ	35
Scrophulariaceæ	28
Rosaceæ	27
Filices	25
Umbelliferæ	24
Cruciferæ	21
	525

Louisiana.	
Compositæ	172
Leguminosæ	70
Gramineæ	66
Cyperaceæ	58
Scrophulariaceæ	29
Rosaceæ	23
Umbelliferæ	20
Labiatæ	19
Onagraceæ	19
Orchidaceæ	18
	494

Colorado.	
Compositæ	161
Leguminosæ	82
Gramineæ	78
Cyperaceæ	53
Scrophulariaceæ	42
Rosaceæ	38
Ranunculaceæ	37
Cruciferæ	30
Saxifragaceæ	28
Polygonaceæ	28
	577

Illinois.	
Compositæ	132
Cyperaceæ	92
Gramineæ	75
Leguminosæ	49
Rosaceæ	36
Scrophulariaceæ	28
Labiatæ	27
Ranunculaceæ	26
Filices	25
Liliaceæ	23
	513

Peoria.	
Compositæ	158
Cyperaceæ	90
Gramineæ	84
Leguminosæ	45
Rosaceæ	30
Labiatæ	27
Ranunculaceæ	26
Scrophulariaceæ	26
Umbelliferæ	20
Filices	20
	526

Washington, D. C.	
Compositæ	130
Cyperaceæ	89
Gramineæ	85
Leguminosæ	45
Rosaceæ	34
Labiatæ	31
Filices	28
Scrophulariaceæ	27
Ericaceæ	25
Ranunculaceæ	24
	518

The relative number of Compositæ increases toward the south and west, and only in Massachusetts and New York they do not take the first place; the Gramineæ take nearly throughout the third place. The number of Leguminosæ increases southward in a great proportion; while the Rosaceæ prevail in the north and decrease southward. The Ericaceæ, so numerous in the east, are little represented in the west; the Labiatæ increase southward, the Scrophulariaceæ westward; the Orchidaceæ and Filices take a higher position in rank in the north than in the south; the Ranunculaceæ are most represented in Colorado, and only little in the south. The Liliaceæ have only in Ohio, Illinois and Wisconsin a place amongst the ten prevalent orders. Umbelliferæ and Cruciferæ increase in number southwestward, and so do the Onagraceæ. In Colorado we find the Saxifragaceæ and Polygonaceæ amongst the prominents, the latter chiefly by the numerous species of Eriogonum, of which sixteen are reported from that state, and the like number inhabit the Rocky Mountains.

The floras of the Northern States (Gray) and Southern States (Chapman) west of the Mississippi compared in the same way show the following figures:

N. Sts ..... Compositæ......... 122	Cyperaceæ........... 104	Gramineæ.............. 75	Leguminosæ 41
S. Sts ..... Compositæ....... 148	Cyperaceæ........... 92	Gramineæ.............. 71	Leguminosæ 54
N. Sts ..... Rosaceæ........... 31	Ericaceæ............. 28	Liliaceæ................. 24	Filices.................... 24
S. Sts ..... Labiatæ............ 27	Scrophulariaceæ.. 25	Ericaceæ 24	Liliaceæ 24
N. Sts ..... Orchidaceæ 23	Scrophulariaceæ.. 22	Ranunculaceæ....... 21	Labiatæ 21
S. Sts ..... Rosaceæ............ 22	Orchidaceæ......... 21	Filices................... 20	Ranunculaceæ...... 19
N. Sts ..... Cruciferæ.......... 19	Umbelliferæ........ 16	Caryophyllaceæ..... 16	Saxifragaceæ........ 15
S. Sts ..... Umbelliferæ....... 17	Euphorbiaceæ 15	Onagraceæ 15	Caryophyllaceæ..... 14
N. Sts ..... Juncaceæ 13	Onagraceæ 12	Euphorbiaceæ....... 12	Caprifoliaceæ 12
S. Sts ..... Cruciferæ.......... 14	Saxifragaceæ....... 13	Polygonaceæ......... 12	Asclepiadaceæ 12
N. Sts ..... Najadaceæ 12	Polygonaceæ 11	Gentianaceæ......... 10	Cupuliferæ.......... 10
S. Sts ..... Convolvulaceæ.... 11	Gentianaceæ........ 10	Cupuliferæ........... 10	Hypericaceæ......... 10

Comparing both rows, we find that the four first orders keep the same rank; that the Labiatæ, Scrophulariaceæ, Euphorbiaceæ, and Onagraceæ take a higher position in the south than in the north; and vice versa the Rosaceæ, Filices, Cruciferæ, and Saxifragaceæ. In the second row we miss the Juncaceæ, Caprifoliaceæ, and Najadaceæ; they are replaced by the Asclepiadaceæ, Convolvulaceæ, and Hypericaceæ, that are not represented in the first row; the rest take nearly the same position in both rows.

GEOGRAPHICAL DISTRIBUTION OF OUR GENERA AND SPECIES.

Those fanciful believers in centres of creation (not centres of preservation as proposed by Bentham), rejecting any theory of descent, set a great value upon the so called endemism and monotypes. As they admit only of recent means of distribution, excluding all geological agencies, endemism is to them the principal proof of an originality of certain floral areas. Monotypes are mostly the arbitrary make of systematists prone to narrow limitations of genera or to wide limitations of species, or the result of an incomplete knowledge of species. Hepatica is a monotype as soon as separated from Anemone and when at the same time we join as varieties with the species H. triloba the little defined species, that have been proposed; but as soon as the latter be acknowledged as species, the genus would cease to be a monotype. Pentachæta was a monotype when Nuttall proposed it, but Gray decribed a second species and joined another monotype Aphantochæta; so both of them ceased to be monotypes. Such examples are many; only from the order compositæ may be mentioned: Cor-

ethrogyne D. C. now with 3 species, Hulsea T. & Gr. with 6, Actinolepis
D. C. after Bentham with 8 species; Oxyura D. C. recently united with
Layia, Tuckermania Nutt with Leptosyne, Coinogyne Less. with Jaumea
Pers.

Of what value endemism is to the believed originality of floral areas
and how little contented we ought to be with exclusively recent agencies
of distribution shows Phryma leptostachya a monotype genus and after
Schauer even a monotype order. It is inimaginable that this plant now
found only in North America and in the Himalayas, by means of recent
agencies could have migrated from one to the other of its actual habitats.
Its disjoint existence is explicable only from geological and climatical
changes and progressive extinction of the plant in the intermediate countries; for a double origin cannot be imagined. It is true paleontologists
made not known this or another related species from former geological
periods, but it is a fact, that other monotypes f. i. Liriodendron now only
found in North America existed in the tertiary period in very distant
countries, in Greenland as well as in Germany and Italy. Should that not
lead to the conclusion that, what are now monotypes, be the last members
of once widely distributed genera now in process of extinction? Analogous
examples are offered in zoology; only compare the small unmber of recent
Ganoids with the richness of former geological periods.

Lately not less than 10 California monotypes were proposed, the greater
part probably waiting for the company of new foundlings.

Systematization so much subject to change united recently Solea to
Ionidium, Zizia to Pimpinella and Gymnostichum with Asprella, so that
our flora would now contain only 21 monotypes, viz: Hydrastes, Sanguinaria, Anychia, Napaea, Floerkea, Apios, Gymnocladus, Echinocystis,
Dodecatheon, Phryma, Jsanthus, Menyanthes, Montelia, Sassafras, Dirca,
Peltandra, Aplectrum, Schollera, Dulichium, Brachyelytrum and Diarrhena.
Caulophyllum and Ieffersonia are excluded; since in addition to each a
second species is known from Eastern Asia (Amur). Besides Phryma only
Menyanthes occurs outside of the American continent: it is widely distributed over Europe and Asia. That was no monotype to Linnæus, when
he proposed the genus, but afterwards there were three genera formed of
it, Limnanthemum and Villarsia; so Menyanthes trifoliata became a monotype. A small area cover Hydrastes, Napæa and Diarrhena; westward
reach the Pacific coast only Floerkea, Dodecatheon and Dirca, the Gulf
coast Sanguinaria, Anychium, Apios, Phryma, Sassafras, Dirca, Peltandra,
Dulichium and Brachyelytrum. Over the Alleghanies do not pass Napæa
and Diarrhena. Westward occur Schollera in Nevada, Montelia in Texas,
and in the western plains Apios, Gymnocladus, Echinocystis, Peltandra (?)
Aplectrum and Dulichium.

Of the 384 genera of our local flora 22 are restricted to the east of
North America. These are, besides the above mentioned monotypes:
Elodes, Boltonia, Blephilia, Onosmodium, Carya, Oryzopsis, Eatonia and

Tricuspis, of the exclusively North American genera reach the Rocky Mountains or the Pacific coast 35: Lechea, Callirhoe, Ptelea, Floerkea, Amorpha, Petalostemon, Apios. Baptisia, Proserpinaca, Heuchera, Gaura, Thaspium, Liatris, Chrysopsis, Polymnia, Silphium, Echinacea, Rudbeckia, Lepachys, Krigia, Troximon, Dodecatheon, Aphyllon, Chelone, Pentstemon, Monarda, Physostegia, Pycnanthemum, Hydrophyllum, Ellisia, Amsonia, Dirca, and Bouteloua.

Of our genera restricted to the Western Hemisphere occur with single species in Mexico or South America 12: Ceanothus, Oenothera, Parthenium, Heliopsis, Helenium, Seymeria, Gerardia, Castilleja, Asclepias, Oxybaphus and Echinodorus, 8 are chiefly South American and partly North American: Asimina, Cuphea, Kuhnia, Actinomeris, Helianthus, Dysodia, Conobea and Tradescantia.

A number of genera are common to North America and Eastern Asia or the Himalayas, of which 26 are represented in our flora amongst which is Phryma. Genera with one species in each continent are: Menispermum, Jeffersonia, Caulophyllum, Podophyllum, Nelumbium, Negundo, Crypotænia, Penthorum, Hamamelis, Saururus, Symplocarpus, and of our two species of Osmorrhiza one is found in Eastern Asia; the 12 species of Dicentra are equally divided, 6 for each continent. Chiefly Asiatic are: Ampelopsis, Amphicarpæa, Hydrangea, Arisæma. Chiefly American: Claytonia, Mitella, Archemora, Triosteum, Nabalus (now united with Prenanthes), Lophanthus, Phlox, Uvularia and Trillium.

Many tropic genera are represented by one or a few species in North America, of which we find in our flora the following:

N.B. * Means that the genus is a large one with 50 and more species, † means that the genus is chiefly South American.

Polanisia.
Mollugo.
*Zanthoxylum.
*Crotalaria.
*Psoralea.
*Tephrosia.
*Desmodium.
*Cassia.
Desmanthus.
*Phaseolus.
*Impatiens.
Ammannia.
Sicyos.
*Spermacoce.
Cephalanthus.
*Vernonia.
†*Eupatorium.
Eclipta.
†Ambrosia.
Erechtites.
†*Lobelia.

Gaylussacia.
*Tecoma.
*Diospyros.
*Plantago.
Ilysanthes.
*Ruellia.
*Verbena.
*Lippia.
†*Ipomoea.
†*Solanum.
*Physalis.
Datura.
*Aristolochia.
Phytolacca.
*Euphorbia.
*Acalypha.
*Laportea.
*Pilea.
Boehmeria.
*Celtis.

Hypoxis.
*Habenaria.
Spiranthes.
Liparis.
*Dioscorea.
Pontederia.
*Cyperus.
Fimbristylis.
*Scleria.
Leersia.
*Vilfa.
*Sporobolus.
*Muhlenbergia.
*Panicum.
Cenchrus.
*Andropogon.
Chrysopogon.
†Adiantum.
Pteris.
Azolla.

All the other genera are either cosmopolitan or predominate in the northern temperate zone of both hemispheres, though some are chiefly North American as: Aster, Erigeron, Solidago, Polemonium, Fraxinus and Comandra.

When the entire upper Mississippi valley to the 38° N. L. emerged the last time, after the drift period, then it is obvious that the whole concern of our flora must have immigrated. The relationship and the present center of preservation of each species will point to the direction of the probable immigration.

There are only a few species, that so far are observed only in the upper Mississippi district: Desmodium Illinoense A.Gr. Aster anomalus Engelm., Phlox bifida Beck., Asclepias Meadii Torr., Trillium recurvatum Beck., Eleocharis Wolfii Gr. These may elsewhere been overlooked or taken for other species; they may have been formed by variation or hybridation in the present period, but not spread over a wider territory or they may have covered a wider area and be restricted now to their present habitat; all that is possible but not proved.

The majority of the species no doubt came from the Alleghanies; for not less than 715 extend to that mountain range and 642 of them surpass the same. A good many of such species, that America has in common with the eastern continent, (114 of our flora) may have immigrated from the north.

All the species that in the same latitude do not reach the Alleghannies but have their eastern limits in Ohio, Indiana, or Illinois may be either western or southern. The southern species (excepted those that keep along the coast) go generally farther northward in the west than in the east of the Alleghannies.

Besides the above mentioned 6 species 48 others do not extend to Ohio, and of those came probably from the south 13:

Nasturtium sessilifolium.
Callirhœ triangulare.
Cornus asperifolia.
Eupatorium serotinum.
Liatris cylindracea.
Rudbeckia subtomentosa.
Coreopsis palmata.

Coreopsis lanceolata.
Lobelia leptostachys.
Amsonia tabernaemontana.
Carya olivæformis.
Leersia lenticularis.
Panicum autumnale.

From southwest 10:

Desmantus brachylobus.
Liatris pycnostachya.
Erigeron divaricatum.
Ambrosia bidentata.
Ambrosia psilostachya.

Dysodia chrysanthemoides.
Hieracium longipilum.
Euphorbia heterophylla.
Echinodorus rostratus.
Cyperus acuminatus.

From west 20:

Clematis Pitcheri.
Anemone decapetala.
Nasturtium sinuatum.
Psoralea floribunda.
Petalostemon violaceum.
Petalostemon candidum.
Amorpha canescens.
Lythrum alatum.
Solidago missouriensis.
Chrysopsis villosa.

Echinacea angustifolia.
Helianthus rigidus.
Androsace occidentalis.
Mimulus Jamesii.
Gerardia grandiflora.
Verbena bracteosa.
Lithospermum angustifolium.
Oxybaphus nyctagineus.
Amarantus blitoides.
Equisetum laevigatum.

From northwest 4:

Viola delphinifolia.
Artemisia ludoviciana.

Troximon cuspidatum.
Stipa spartea.

From north 1:

Equisetum palustre.

There are 49 species that extend eastward only to Ohio. Those marked * occur in the southern Alleghannies, but do not reach those mountains in the latitude of Illinois.

From southeast may come 5:

*Aster Shortii.
Solidago Riddelhi.
*Silphium perfoliatum.

*Prenanthes crepidinea.
*Fraxinus quadrangulata.

From south 20:

Isopyrum biternatum.
Thelypodium pinnatifidum.
Hypericum sphaerocarpum.
*Ptelea trifoliata.
*Baptisia leucantha.
*Baptisia leucophaea.
Spermacoce glabra.
*Vernonia fasciculata.
Silphium laciniatum.
*Silphium terebinthaceum.

Coreopsis aristosa.
Cacalia tuberosa.
*Ruellia ciliosa.
Verbena stricta.
Onosmodium molle.
Lithospermum latifolium.
*Gentiana pubera.
Acerates longifolia.
Carex crus corvi.
Eragrostis pectinacea.

From southwest 10:

Ammannia coccinea.
Silphium integrifolium.
Helianthus grosseserratus.
Verbesina helianthoides.
Prenanthes aspera.

Conobea multifida.
Seymeria macrophylla.
Asclepias Sullivantii.
Diarhena americana.
Montelia tamariscina.

From west 4:

Prenanthes racemosa.
Cuscuta glomerata.

Scilla Fraseri.
Equisetum robustum.

From northwest 8:

Ludwigia polycarpa.
Aster junceus.
Artemisia biennis.
Habenaria leucophaea.

Trillium nivale.
Eragrostis Frankii.
Hordeum pratense.
Hordum jubatum.

From north 2:

Carex arida.

Poa sylvestris.

Of 50 species, that reach the Alleghannies in the latitude of Illinois, but do not pass over, are probably southern 11:

Asimina triloba.
Gleditschia triacanthos.
Amorpha fruticosa.
Boltonia asteroides.
Eclipta alba.
Rudbeckia triloba.

Lepachys pinnata.
Coreopsis tripteris.
Ruellia strepens.
Ipomoea lacunosa.
Euphorbia commutata.

From southwest 4:

Aster sericeus.
Parthenium integrifolium.

Lippia lanceolata.
Euphorbia dentata.

From west 2:

Helianthus laetiflorus.

Melica mutica.

From northwest 1:

Spiraea Aruncus.

The rest (32 species) extend from the Alleghannies westward, but not eastward.

Delphinium tricorne.
Jeffersonia diphylla.
Napaea dioica.
Hibiscus militaris.
Rhamnus lanceolatus.
Aesculus glabra.
Trifolium reflexum.
Psoralea onobrychis.
Desmodium sessilifolium.
Gymnocladus canadensis.
Spiraea lobata.
Agrimonia parviflora.
Rosa Setigera.
Heuchera hispida.
Galium concinnum.
Eupatorium altissimum.

Aster azureus.
Solidago Ohioensis.
Polymnia canadensis.
Echinacea purpurea.
Helianthus doronicoides.
Helianthus occidentalis.
Cacalia reniformis.
Cacalia atriplicifolia.
Steironema longifolia.
Pycnanthemum pilosum.
Scutellaria versicolor.
Phlox maculata.
Phlox glaberrima.
Ellisia nyctelaea.
Cuscuta inflexa.
Cypripedium candidum.

All the other species we have in common with the Eastern States; but judging from the centres of preservation, it seems that we have from the South 20.

Polygala incarnata.
Crotalaria sagittalis.
Desmodium paucillorum.
Cercis canadensis.
Cassia marilandica.
Cassia chamaecrista.
Mollugo verticillata.
Erynchium yuccaefolium.
Rudbeckia hirta.
Diospyros virginiana.

Tecoma radicans.
Verbena hastata.
Verbena urticifolia.
Lithospermum hirtum.
Ipomoea pandurata.
Solanum carolinense.
Euphorbia corollata.
Croton glandulosus.
Cyperus erythrorhizus.
Azolla caroliniana.

From Southwest 9.

Linum sulcatum.
Kuhnia eupatorioides.
Monarda fistulosa.
Physostegia virginiana.
Cuscuta tenuiflora.

Polygonum tenue.
Euphorbia dentata.
Cyperus phymatodes.
Hemicarpha subsquarrosa.

From West 8.

Ranunculus fascicularis.
Sisymbrium canescens.
Ceanothus americanus.
Negundo aceroides.

Salix longifolia.
Eatonia obtusata.
Festuca tenella.
Carex umbellata.

From Northwest 44.

Actaea alba.
Claytonia Virginica.
Ramnus alnifolius.
Astragalus canadensis.
Potentilla norvegica.
Potentilla arguta.
Rubus occidentalis.
Rosa blanda.
Epilobium palustre.
Epilobium coloratum.
Heracleum lanatum.
Osmorhiza longistylis.
Osmorhiza brevistylis.
Aralia racemosa.
Aralia nudicaulis.
Cornus stolonifera.
Galium Aparine.
Dodecatheon Meadia,
Utricularia vulgaris.,
Veronica Anagallis.
Veronica Americana.
Veronica peregrina.

Mentha canadensis.
Scutellaria galericulata.
Lithospermum canescens.
Polemonium reptans.
Menyanthes trifoliata.
Ceratophyllum demersum.
Populus tremuloides.
Typha latifolia.
Alisma Plantago.
Smilacina stellata.
Carex lanuginosa.
Carex Richardsoni.
Carex Buxbaumii.
Carex stricta.
Carex disticha.
Carex teretiuscula.
Carex stellulata.
Calamagrotis canadensis.
Köleria cristata.
Poa pratensis.
Poa serotina.
Elymus striatus.

From North 23.

Viburnum Opulus.
Achillea millefolium.
Senecio aurea.
Taraxacum dens leonis.
Arctostaphyllos uva ursi.
Lysimachia thyrsiflora.
Utricularia intermedia.
Veronica scutellata.
Lycopus sinuatus.
Stachys palustris.
Polygonum aviculare.
Salix candida.

Salix cordata.
Salix myrtilloides.
Sparganium eurycarpum.
Triglochin palustre.
Triglochin maritimum.
Habenaria hyperborea.
Eleocharis palustris.
Eriophorum gracile.
Rynchospora alba.
Carex limosa.
Poa compressa.

In common with the Eastern Continent, our flora contains 114 species; of these six (Ranunculus multifidus, Claytonia virginica, Mitella diphylla, Artemisia biennis, Bromus ciliatus and Adiantum pedatum) extend only to East Siberia and not farther westward.

Of the 116 species, which, after A. de Candolle (Géographie Botanique, p. 564), are widely distributed, our flora contains thirty-two, all European, except Eclipta alba, a tropical plant, and Erigeron canadense, an immigrant in Europe.

Species that are found in Australia (Hooker's Introduction to the Flora of Australia) we have twenty-five, in Guyana (Schomburgk) are twenty-three, and Grisebach's West Indian plants contain fifty of our flora.

Only three species extend to the polar region of Northern Greenland: Habenaria hyperborea, Calamagrostis canadensis and Bromus Kalmii. Besides those three forty-six more extend to the artic circle.

The distribution of our species over the natural floral districts of N. America will show in the following tables:

DISTRIBUTION OF THE SPECIES OF THE FLORA OF PEORIA OVER NORTH AMERICA.

This table shows how many species of each order the different floral provinces have in common with Peoria:

	Peoria	Can.	N. Atl.	All.	Ohio.	S. Atl.	Jan.	Up Mo.	R. Mts.	N. Mex.	Nev.	Cal.	Ore.	Huds.	Alaska	Aret.
Ranunculaceae	22	18	18	19	20	14	17	18	13	10	5	1	1	10	4	2
Anonaceae	1	1	...	1	1	1	1									
Menispermaceae	1	1	1	1	1	1	1	1								
Berberidaceae	3	2	2	3	3	1	1	1								
Nymphaeaceae	3	3	3	3	3	3	3	1	1	1	1	...	1	1		
Papaveraceae	1	1	1	1	1	1	1									
Fumariaceae	2	2	2	2	2	1	1	1	1	1	1	...	1			
Cruciferae	14	11	11	11	12	8	12	6	5	4	5	4	4	3	2	3
Capparidaceae	1	1	1	1	1	1	1									
Violaceae	6	4	5	5	5	3	3	3	2	...	1	1	1	2	...	1
Cistaceae	3	3	3	3	3	3	3	1								
Hypericaceae	5	4	4	4	5	2	3	1								
Caryophyllaceae	7	6	6	7	7	3	5	5	5	3	3	2	3	2	3	2
Paronychieae	1	1	1	1	1	1	1	1								
Portulacaceae	1	1	1	1	1	1	1	1	1	1						
Malvaceae	3	...		3	2	...	2									
Tiliaceae	1	1	1	1	1	...	1						1			
Linaceae	1	1	1	1	1	...		1								
Geraniaceae	6	6	6	6	6	5	5	6	5	2	1	1	3	1		
Rutaceae	2	2	1	1	2	2	2	2	1	1						
Anacardiaceae	3	3	3	3	3	3	3	3	2	2	3	1	2			
Vitaceae	4	4	4	4	4	3	3	3	3	3						
Rhamnaceae	3	2	2	3	3	1	2	1	1	1	...	1	...	1		
Celastraceae	2	2	2	2	2	1	2	2								
Sapindaceae	5	4	4	5	5	3	4	2	1	1	1	...	1	1		
Polygalaceae	4	4	4	4	4	3	4	2	1							
Leguminosae	38	29	24	30	32	28	33	30	7	13	2	2	3	5	1	2
Rosaceae	25	23	21	24	24	13	18	15	10	8	5	3	3	7	2	3
Saxifragaceae	7	6	6	7	7	2	3	2	1	...				1		
Crassulaceae	1	1	1	1	1	1	1									
Hamameliaceae	1	1	1	1	1	1	1									
Haloragene	1	1	1	1	1	1	1		...	1						
Onagraceae	10	8	8	8	9	4	7	10	6	3	1	3	5	3	1	1
Lythraceae	4	2	2	2	3	2	4	2	2	2	1	2				
Cucurbitaceae	2	2	2	2	2	1	1	1	1	...			1			
Ficoideae	1	1	1	1	1	1	1	1	1	1	1	1				
Umbelliferae	17	17	17	17	17	8	11	8	7	4	5	4	2	3	1	
Araliaceae	3	3	3	3	3	...	1	2	2	...				2		
Cornaceae	6	5	5	6	5	4	4	1	1	1			1	1	1	1
Caprifoliaceae	8	8	8	8	8	2	3	2	3	...				4	...	1
Rubiaceae	7	5	5	6	7	5	5	4	4	4	3	4	3	3	2	1
Compositae	131	97	79	101	112	68	97	81	41	42	18	15	12	32	4	6
Lobeliaceae	6	5	5	5	5	3	6	4	1	1						
Campanulaceae	3	3	3	3	3	2	3	1	1	1	1	1	1			
Ericaceae	4	3	4	4	4	2	1	2	2	...	1	2	2	1	1	1
Ebenaceae	1	...	1	1	1	1	1	1								
Primulaceae	7	5	4	6	6	2	4	4	3	1	4	2	2	1	1	1
Oleaceae	5	5	4	4	5	3	3	2	...	1	1	...	2			
Apocynaceae	3	2	2	2	2	3	3	3	2	2	2	2	2	1		
Asclepiadaceae	12	9	9	9	11	6	8	6	3	5						
Gentianaceae	5	4	4	4	5	1	1	1	1	1	1	1	1	1	1	1
Polemoniaceae	6	3	3	5	5	3	4	1								
Hydrophyllaceae	3	2	2	3	3	1	2	3	2	...			1			
Borraginaceae	8	6	5	5	7	3	5	1	3	2	1	1	1	2	...	

	Peoria.	Can...	N. Atl...	All...	Ohio...	S. Atl...	La...	Up Mo.	R. Mts...	N. Mex	Nev...	Cal...	Ore...	Huds...	Alaska	Arct...
Convolvulaceae	10	5	7	9	10	6	7	3	3	5	2	1	1	1	...	...
Solanaceae	5	5	5	5	5	5	5	5	4	2	2	2	2	...	...	
Scrophulariaceae	22	17	17	17	19	11	16	10	7	6	5	5	7	3	2	...
Orobanchaceae	1	1	1	1	1	1	1	1	1	1	1	1	1	...	...	
Lentibulariaceae	2	2	2	2	2	...	1	1	1	...	1	1	1	2	...	
Bignoniaceae	1	...	...	1	1	1	1	...	...							
Acanthaceae	2	2	...	2	2	2	2	1	...	1	...					
Verbenaceae	6	3	3	4	5	4	6	5	4	5	2	2	2	...	...	
Labiatae	23	19	21	23	23	11	19	10	9	7	8	5	5	6	1	...
Plantaginaceae	3	2	3	3	3	2	3	1	...	1	...	1	...	...	...	
Aristolochiaceae	2	2	2	2	2	1	2	1	...							
Nyctaginaceae	1	...	...	...	...	...	1	1	1	1	...					
Phytolaccaceae	1	1	1	1	1	1	1	1	...	1	...					
Chenopodiaceae	2	2	2	2	2	2	2	2	2	2	2	2	2	2	...	
Amarantaceae	4	3	2	2	3	2	3	4	3	4	2	2	2	...	...	
Polygonaceae	16	16	16	16	16	11	13	10	5	8	4	4	4	4	1	1
Lauraceae	1	1	1	1	1	1	1	...	...							
Thymelaceae	1	1	1	1	1	1	1	1	1	...	1	...				
Santalaceae	1	1	1	1	1	...	1	1	1	1	...	1	1	...	...	
Saururaceae	1	1	1	1	1	1	...									
Ceratophyllaceae	1	1	1	1	1	1	1	1	1	1	1	1	...	1	...	
Callitrichaceae	1	1	1	1	1	1	1	1	1	1	...	1	1	1	1	1
Euphorbiaceae	9	5	4	8	8	7	7	6	1	6	1	1	1	...	...	
Urticaceae	10	10	10	10	10	7	10	10	3	7	4	1	1	...	...	
Plantanaceae	1	1	1	1	1	1	1	...	1	...						
Juglandaceae	8	7	6	7	7	4	7	...	...	2	...					
Cupuliferae	12	11	11	11	12	9	11	5	...	3	...					
Salicaceae	13	13	12	12	13	3	4	5	7	5	4	1	1	2	2	3
Coniferae	2	2	1	2	2	1	1	1	1	1	1	1	1	2	...	1
Araceae	5	5	5	5	5	5	4	1	...							
Lemnaceae	4	3	4	4	4	2	3	3	3	3	3	2	2	3	...	
Typhaceae	2	2	2	2	2	1	1	2	1	1	2	1	2	2	...	1
Najadaceae	6	6	6	6	6	5	5	4	3	5	4	3	3	4	...	...
Alismaceae	7	6	6	6	6	4	3	5	4	2	4	3	3	4	2	2
Hydrocharidaceae	2	2	2	2	2	2	1	...	...	1	...					
Orchidaceae	16	16	14	15	16	7	8	5	5	1	1	...	2	1	1	1
Amaryllidaceae	1	1	1	1	1	1	1	1	1	...						
Iridaceae	2	2	2	2	2	2	2	2	...	1	1	1	1	...	1	...
Dioscoreaceae	1	1	1	1	1	1	1	1	...							
Smilacaceae	2	2	2	2	2	1	2	1	...	1	1	...				
Liliaceae	13	12	10	10	12	3	7	7	4	3	3	2	3	3	...	1
Juncaceae	4	4	4	4	4	2	3	2	2	2	1	2	2	2	...	
Pontederaceae	2	2	2	2	2	2	2	1	...	1	1	...				
Commelynaceae	2	1	1	1	1	2	2	1	...	1	1	...				
Cyperaceae	78	70	66	74	76	43	41	38	20	26	10	13	15	23	5	3
Gramineae	70	63	56	61	66	41	45	41	33	20	23	17	6	11	6	5
Equisetaceae	7	5	4	5	6	1	1	1	5	4	3	3	3	1	1	2
Filices	16	15	16	16	16	15	11	7	5	3	4	5	5	4	5	2
Ophioglossaceae	1	1	1	1	1	1	1	1	1	1	1	1	1	1	...	
Lycopodiaceae	1	1	1	1	1	1	1	...	...	1	...					
Hydropterides	1	1	1	1	1	1	1	...	...	1	...					
Dicotyledoneae	592	479	450	512	543	334	443	338	201	202	108	88	88	118	31	32
Monocotyledoneae	217	198	184	198	208	124	130	114	76	66	54	44	39	54	15	13
Cryptogamae vasculares	26	23	23	24	25	10	15	14	11	9	8	9	9	6	6	4
	835	700	657	734	776	477	588	466	288	277	170	141	136	178	52	49

PLANTS OF THE ILLINOIS FLORA, BUT SO FAR NOT FOUND NEAR PEORIA.

[S.--Southward. N.—Northward, W.—Westward.

Ranunculaceæ.
S.   Clematis viorna L.
N.   Anemone patens L. var. Nuttalliana.
N.   A. nemorosa L.
      Ranunculus divaricatus Schrank.
      R. aqualilis L. var. trichophyllus.
S.   R. alismaefolius Geyer.
S.   R. oblongifolius Ell.
N.   R. rhomboideus Goldie.
      R. pennsylvanicus L.
S.   Myosurus minimus L.
S.   Trautvetteria palmata Fisch. and Mey.
W.   Delphinium azureum Michx.
S.   Cimicifuga racemosa Ell.

Magnoliaceæ.
S.   Liriodendron tulipifera L.

Nymphaeaceæ.
      Brasenia peltata Ph.

Sarraceniaceæ.
N.   Sarracenia purpurea L.

Papaveraceæ.
S.   Stylophorum diphyllum Nutt.

Fumariaceæ.
N.   Adlumia cirrhosa Raf.
N.   Dicentra canadensis D. C.
N.   Corydalis glauca Ph.

Cruciferæ.
S.   Nasturtium obtusum Nutt.
N.   Arabis lyrata L.
      A. hirsuta Scop.
N.   A. perfoliata Lam.
N.   A. Drummondii Gr.
S.   A. Ludoviciana Mey.
N.   Barbarea vulgaris R. Br.
      Erysimum cheiranthoides L.
      E. asperum D. C. var. Arkansanum.
S.   Draba cuneifolia Nutt.
      Lepidium intermedium Gr.

Violaceæ.
N.   Viola lanceolata L.
N.   V. blanda Willd.
      V. canina L. var. sylvestris.
      V. striata Ait.
      V. tricolor L. var arvensis.

Cistaceæ.
N.   Hudsonia tomentosa Nutt.

Hypericaceæ.
S.   Ascyrum crux Andreae L.
S.   H. proliferum L.
S.   H. adpressum Bart.
S.   H. ellipticum Hook.
S.   H. Drummondii T. Gr.
      H. mutilum L.
      H. canadense L.
      H. sarothra Michx.

Elatinaceæ.
S.   Elatine americana Arnott.
S.   Bergia Texana Seubert.

Caryophyllaceæ.
      Silene virginica L.
S.   S. regia Sims.
S.   Sagina apetala L.
N.   Stellaria crassifolia Ehrh.
N.   Arenaria stricta Michx.
      Cerastium oblongifolium Torr.

Portulacaceæ.
W.   Claytonia caroliniana Michx.
N.   Talinum teretifolium Ph.

Malvaceæ.
S.   Hibiscus Moscheutos L.
S.   H. grandiflorus Michx.
N.   Sphaeralcea acerifolia Nutt.

Tiliaceæ.
S,   Til'a heterophylla Vent.

Linaceæ.
      Linum virginianum L.
S.   L. striatum Walt.

Geraniaceæ.
      Geranium carolinianum L.

Ilicineæ.
      Ilex verticillata Gr.
S.   I. decidua Walt.

Anacardiaceæ.
      Rhus typhina L.
      R. copallina L.
N.   R. venenata DC.

Vitaceæ.
S. Vitis indivisa Willd.
S. V. bipinnata TGr.

Rhamnaceæ.
S. Frangula caroliniana Gr.
N. Ceanothus ovalis Big.

Celastraceæ.
Euonymus americanus L.

Sapindaceæ.
S. Acer rubrum L.

Polygalaceæ.
Polygala ambigua Nutt.
P. polygama Walt.
N. P. cruciata L.
N. P. pauciflora Willd.

Leguminosæ.
N. Lupinus perennis L.
Trifolium stoloniferum Muhl.
N. Petalostemon foliosus Gr.
S. Psoralea melilotoides Michx.
S. Dalea alopecuroides Willd.
S. Robinia pseudacacia L.
S. Wistaria frutescens DC.
S. Astragalus mexicanus ADC.
S. A. distortus TGr.
N. A. platensis Nutt var Tennessee-
nsis.
S, Desmodium rotundifolium DC.
S. D. laevigatum DC.
S. D. viridiflorum Beck.
S. D. rigidum DC.
S. D. ciliare DC.
S. D. marilandicum DC.
S. Lespedeza repens TGr.
S. L. hirta Ell
S. Stylosanthes elatior Sw.
S. Phaseolus perennis Walt.
Ph. pauciflorus Benth.
S. Clitoria Mariana L.
S. Galactia mollis Michx.
N. Vicia caroliniana Walt.
N. Lathyrus venosus Muhl.
N. L. ochroleucus Hook.
N. Baptisia tinctoria RBr.
S. Cassia nictitans L.
S. Gleditschia monosperma Walt.

Rosaceæ.
N. Prunus pumila L.
N. P. pennsylvanica L.
Neillia opulifolia Benth, Hook.
Spiræa salicifolia Raf.

N. Sp. tomentosa L.
Geum strictum Ait.
S. G. vernum TGr.
N. G. macrophyllum Willd.
N. G. rivale L.
N. G. triflorum Ph.
Fragaria vesca L.
S. Gillenia stipulacea Nutt.
S. Poterium canadense Gr.
S. Potentilla paradoxa Nutt.
N. P. fruticosa L.
N. P. palustris L.
Rubus strigosus Michx.
R. canadensis L.
N. R. triflorus Rich.
N. R. hispidus L.
S. Crataegus arborescens Ell.
S. Pyrus angustifolia Ait.
N. P. arbutifolia L.

Saxifragaceæ.
N. Ribes cynosbati L.
N. R. hirtellum Michx.
S. Saxifraga Forbesii Vasey.
S. Heuchera americana L.
S. H. Rugelii Shuttl.

Crassulaceae.
Sedum ternatum Michx.
S. S. pulchellum Michx.

Hamameliaceae.
S. Liquidambar styraciflua L.

Halorageae.
Myriophyllum verticillatum L.
M. heterophyllum Michx.
M. scabratum Michx.
N. M. spicatum L,
N. Hippuris vulgaris L.

Onagraceaa.
N. Epilobium angustifolium L.
E. molle Torr.
S. Oenothera linearis Michx.
S. Oe. missouriensis Sims.
Ludwigia sphaerocarpa Ell.

Melastomaceae.
Rhexia virginica L.

Lythraceae.
Peplis diandra Nutt.
S. Decodon verticillatus Ell.

Loasaceae.
S. Mentzelia oligosperma Nutt.

Cactaceae.
Opuntia Rafinesquii Engelm.

Passifloreae.
S. Passiflora lutea L.

Umbelliferae.
Sanicula canadensis L.
Polytaenia Nuttallii DC.
S. Discopleura Nuttallii DC
N. Conioselinum canadense T. Gr.
Eulophus americanus Nutt.
S. Archangelica hirsuta T. Gr.
S. Erigenia bulbosa Nutt.

Araliaceae.
S. Aralia spinosa L.

Cornaceae.
N. Cornus canadensis L. *
S. C. florida L.
S. Nyssa multiflora Wang.

Caprifoliaceae.
Symphoricarpus occidentalis R.Br.
S. S. vulgaris Michx.
N. Diervilla trifida Moench.
N. Viburnum pubescens Ph.
N. V. acerifolium L.
S. Triosteum angustifolium L.

Rubiaceae.
Galium asprellum Michx.
S. G. pilosum Ait.
N. G. boreale L.
S. Diodia virginica L.
S. D. teres Walt.
S. Mitchella repens L.
Houstonia coerulea L.
S. H. angustifolia Michx.
S. H. minima Beck.

Valerianaceae.
Fedia radiata Michx.
N. F. umbilicata Sull.
N. Valeriana edulis Nutt.
S. V. pauciflora Michx.

Compositae.
W. Vernonia Noveboracensis Willd.
S. Elephantopus carolinianus Willd.
Liatris squarrosa Willd.
N. L. spicata Willd.
S. Eupatorium aromaticum L.
S. Eu. coelestinum L.
S. Aster patens Ait.
S. A. turbinellus Lindl.
A. undulatus L.
A. dumosus L.
N. A. acuminatus Michx.
A. ptarmicoides TGr.

S. Boltonia diffusa L'Her.
Solidago bicolor L.
S. caesia L.
N. S. stricta Ait.
S. S. petiolaris Ait.
S. altissima L.
S. S. Drummondii TGr.
S. Radula Nutt.
S. Pluchea foetida DC.
S. Polymnia Uvedalia L.
S. Chrysogonum virginianum L.
S. Rudbeckia speciosa Wender.
S. Helianthus atrorubens L.
S. H. mollis Lam.
S. H. microcephalus TGr.
S. Coreopsis auriculata L.
C. trichosperma Michx.
S. C. discoidea TGr.
N. Bidens cernua L.
B, Beckii Torr.
S. B, bipinnata L.
Verbesine virginica L.
S. Hymenopappus scabiosaeus L'Her.
N. Actinella scaposa Nutt.
S. Leptopoda brachypoda TGr.
S. Matricaria discoidea DC.
N. Artemis.a dracunculoides Ph.
N. A, serrata Nutt.
S. Senecio lobatus Pers.
S. Cnicus virginianus var filipendulum Ph.
S. Krigia virginica Don.
S. K, Dandelion Nutt.
Hieracium Gronovii L.
N. H, canadense Michx.
Prenanthes altissimus Hook.
Lactuca leucophaea Gr.

Lobeliaceae,
S. Lobelia puberula Michx.

Campanulaceæ,
Campanula rotundifolia L.
S. C, divaricata Michx.

Ericaceæ,
N. Vaccinium macrocarpum Ait.
N. V, pennsylvanicum Lam.
N. V, canadense Kalm.
N. V, corymbosum L.
S. V, arboreum Marsh.
N. Andromeda polifolia L.
N. Pyrola elliptica Nutt.
N. P, chlorantha Sw.
Chimaphila umbellata Nutt.

Primulaceæ,
    Lysimachia stricta Ait.
S. Centunculus minimus L.

Sapotaceæ,
S. Bumelia lanuginosa Pers.

Oleaceæ.
S. Forestiera acuminata Poir

Asclepiadaceæ.
S. Asclepias perennis Walt.
    A. variegata L.
N. A. ovalifolia Descaine.
N. Acerates lanuginosa Descaine.
    Asclepiodora viridis Gr.
S. Enslenia albida Nutt

Loganiaceæ.
S. Spigelia marilandica L.

Gentianaceæ.
N. Gentiana crinita Froel.
N. G. detonsa Walt.
N. G. saponaria L.
S. Sabbatia angularis Ph.
N. Bartonia tenella Muhl.
S. Frasera carolinensis Walt.

Polemoniaceæ.
S. Phlox paniculata L.
S. Ph. stellaria Gr.

Hydrophyllaceæ.
    Hydrophyllum canadense L.
S. Ph. bipinnatifida Michx.

Borraginaceæ.
S. Cynoglossum virginicum L.
S. Heliotropium curassavicum L.

Convolvulaceæ.
W. Breweria Pickeringii Gr.
S. Cuscuta decora Chois.

Solanaceæ.
    Physalis pubescens L.
    Ph. philadelphica Lam.
S. Ph. angulata L.

Scrophulariaceæ.
    Linaria canadensis Spr.
    Collinsia verna Nutt.
    Chelone obliqua L.
    Pentstemon laevigatus Sol. var.
      digitalis Gr.
W. P. grandiflorus Fraser.
S. Mimulus alatus Ait.
S. Herpestes rotundifolia Ph.

S. Gratiola sphaerocarpa Ell.
    Synthyris Houghtoniana Benth.
S. Veronica serpyllifolia L.
S. Buchnera americana L.
    Gerardia Skinneriana Wood.
    G. quercifolia Ph.
    G. pedicularis L.
S. G. flava L.
S. G. laevigata Raf.
N. Castilleja sessiliflora Ph.

Orobanchaceæ.
S. Conopholis americana Wallr.
S. Aphyllon ludovicianum Gr.

Lentibularjaceæ.
    Utricularia biflora Lam.
· U. gibba L.
N. U. minor L.

Bignoniaceæ.
S. Bignonia capreolata L.

Acanthaceæ.
    Dianthera americana L.

Verbenaceæ.
    Verbena angustifolia Michx.

Labiatæ.
    Trichostema dichotomum L.
S. Cunila Mariana L.
S. Pycnanthemum incanum Michx.
N. Calamintha Nuttallii Benth.
    Hedeoma hispida Ph.
S. Collinsonia canadensis L.
S. Salvia lyrata L.
S. Monarda Bradburiana Beck.
    M. punctata L.
    Blephilia ciliata Raf.
S. Synandra grandiflora Nutt.
S. Scutellaria canescens Nutt.
    Stachys cordata Ridd.
    St. hyssopifolia Michx., var am-
      bigua Gr.

Plantaginaceœ,
S. Plantago pusilla Nutt.
S. P. patagonica Jacq., var aristata.

Aristolochiaceæ,
S. Aristolochia tomentosa Sims.

Chenopodiaceæ,
W. Cycloloma platyphyllum Mocq.
N. Blitum capitatum L.
    Atriplex patula L.

Amarantaceæ,
S. Froelichia floridana Mocq.

Polygonaceæ,
   Polygonum Hartwrightii Gr.
   P. articulatum L.
   P. arifolium L.
Lauraceæ,
   Lindera Benzoin Meisner.
Loranthaceæ,
S. Phoradendron flavescens Nutt.
Callitrichaceæ,
S. Callitriche Austini Engelm.
S. C. auctumnalis L.
Euphorbiaceæ.
   Euphorbia Geyeri Engelm.
   Eu. glyptosperma Engelm.
S. Eu. serpens HBK.
S. Eu. humistrata Engelm.
S. Croton capitatus Michx.
S. Cr. monanthogynus Michx.
S. Crotonopsis linearis Michx.
S. Phyllanthus carolinensis Walt.
Urticaceæ.
N. Ulmus racemosa Thomas.
S. U. alata Michx.
S. Celtis mississippiensis Bosc.
Cupuliferæ.
   Quercus stellata Wang.
   Qu. palustris DuRoi.
S. Qu. falcata Michx.
Betulaceæ.
   Alnus serrulata Ait.
   Betula nigra L.
N. B. papyracea.
N. B. lenta L.
N. B. pumila L.
Myricaceæ,
N, Comptonia asplenifolia Ait.
Salicaceæ,
   Salix lucida Muhl.
N. S. rostrata Rich.
S. Populus heterophylla L.
Coniferæ,
N. Pinus Banksiana Lamb.
N. P. strobus L.
S. P. mitis Michx.
N. Larix americana Michx.
N. Juniperus communis L.
N. Taxus baccata L., var canadensis.
Lemnaceæ,
S. Wolffia Brasiliensis Weddell.
Typhaceæ,
N. Sparganium simplex Hudson.

Najadaceæ,
S. Potamogeton Claytoni Tuck.
S. P. pulcher Tuck.
N. P. Vaseyi Robbins.
N. P. spirillus Tuck.
   P. hybridus Michx.
N. P. rufescens Schrad.
N. P. Lonchites Tuck.
   P. gramineus L.
   P. lucens L.
N. P. perfoliatus L.
   P. compressus L.
Alismaceæ,
   Scheuchzeria palustris L.
   Sagittaria graminea Michx.
S. Echinodorus parvulus Engelm.
S. E. radicans Engelm.
Orchidaceæ,
N. Habenaria viridis Spr., var bracteata.
   H. psycodes Gr.
S. H. peramoena Gr.
N. Goodyera pubescens RBr.
S. Spiranthes latifolia Torr.
N. Pogonia ophioglossoides Nutt.
N. Microstylis monophyllos Lindl.
S. M. ophioglossoides Nutt.
S. Corallorhiza multiflora Nutt.
Amaryllidaceæ
S. Pancratium rotatum Ker.
S. Agave virginica L.
Haemodoraceæ,
   Aletris farinosa L.
Smilacaceæ,
S. Smilax rotundifolia L.
S. S. glauca Walt.
S. S. tamnoides L.
Liliaceæ,
   Trillium sessile L.
   Tr. grandiflorum Salisb.
N. Tr. cernuum L.
   Melanthium virginicum L.
N. Zygadenus glaucus Nutt.
S. Stenanthium angustifolium Gr.
W. Veratrum Woodii Robbins.
   Chamaelirium luteum Gr.
N. Tofjeldia glutinosa Willd.
   Uvularia perfoliata L.
N. Smilacina trifolia Desf.
N. S. bifolia Ker.
   Polygonatum biflorum Ell.
   Lilium canadense L.

N. Erythronium americanum Sm.
N. Allium cernuum Roth
W. A. stellatum Nutt.
S. A. striatum Jacq.

Juncaceæ.

Luzula campestris DC.
Juncus effusus L.
J. marginatus Rostk.
J. bufonius L.
N. J. Greenii Oakes & Tuck.
N. J. Vaseyi Engelm.
N. J. alpinus var insignis Engelm.
S. J. acuminatus var robustus Engelm.
S. J. acuminatus var debilis Engelm.
J. brachycarpus Engelm.
N. J. canadensis var longicaudatus Engelm.
N. J. canadensis var coarctatus Engelm.

Commelynaceæ.

S. Commelyna virginica L.
S. Tradescantia pilosa Lehm.

Xyridaceæ,

Xyris flexuosa Muhl.

Cyperaceæ,

S. Cyperus Engelmanni Steud.
C. Schweinitzii Torr.
S. Kyllingia pumila Michx.
S. Eleocharis quadrangulata RBr.
S. E. rostellata Torr.
S. E. Engelmanni Steud.
W. E. Engelmanni var detonsa Gr.
N. Scirpus pauciflorus Lightf.
N. S. caespitosus L.
S. S. debilis Pursh.
S. S. supinus var Hallii Gr.
S. fluviatilis Gr.
S. S. polyphyllus Vahl.
S. Eriophorum Michx.
N. Eriophorum polystachyon L.
Fimbristylis spadicea var castanea Gr.
F. laxa Vahl.
F. capillaris Gr.
Rhynchospora cymosa Nutt.
R. capillacea Torr.
N. R. glomerata Vahl.
N. Cladium mariscoides Torr.
Carex siccata Dew.
C. decomposita Muhl.
C. Muhlenbergii Schk.

S. C. retroflexa Muhl.
N. C. chordorhiza Ehrb.
N. C. Bebbii Olney.
S. C. foenea Willd.
N. C. aquatilis Wahl.
C. aperta Boot.
C. crinita Lam.
N. C. panicea var Bebbii Olney.
C. tetanica Schk.
C. conoidea Schk.
S. C. virescens Muhl.
S. C. plantaginea Lam.
S. C. Careyana Torr.
N. C. platyphylla Carey.
N. C. retrocurva Dew.
N. C. pedunculata Muhl.
S. C. debilis Michx.
N. C. Œderi Ehrb.
S. C. stenolepis Terr.
C. retrorsa Schw.
C. utriculata Boot.
N. C. Tuckermani Boot.
C. bullata Schk,

Gramineæ,

Vilfa virginica Beauv.
N. Sporobolus cryptandrus Gr.
Ammophila longifolia Benth.
S. Aristida dichtoma Michx.
S. A. ramosissima Engelm.
S. A. gracilis Ell.
S. A. stricta Michx.
A. oligantha Michx.
A. purpurascens Poir.
A. tuberculosa Nutt.
W. Bouteloua hirsuta Lag.
N. B. oligostachya Torr.
S. Diplachne fascicularis Benth.
Triplasis purpurea Chap.
N. Glyceria canadensis Trin.
W. Poa alsodes Gr.
Eragrostis tenuis Gr.
S. Uniola latifolia Michx.
S. Lepturus paniculatus Nutt.
N. Triticum repens L.
N. T. caninum L.
Trisetum palustre L.
N. Hierochloa borealis R Sch.
N. Milium effusum L.
S. Paspalum fluitans Kth.
S. P. walterianum Schult.
P. setaceum Michx.
S. P. laeve, Michx.
Panicum filiforme L.

S. F. microcarpon Muhl.
S. Tripsacum dactyloides L.
S. Erianthus saccharoides Michx.
Andropogon dissitiflorus Michx.

Equisetaceæ.
N. Equisetum scirpoides Michx.

Filices.
Polypodium vulgare L.
S. P. incanum Sw.
S. Cheilanthes vestita Sw.
W. Ch. lanuginosa Nutt.
N. Pellaea gracilis, Hook.
P. atropurpurea Link.
S. Asplenium pinnatifidum Nutt.

S. A. Trichomanes L.
A. ebeneum Ait.
S. Phegopteris polypodioides Fee.
Aspidium Noveboracense Sw.
A. Goldianum Hook.
A. marginale Sw.
S. Woodsia obtusa Torr.
N. W. Ilvesis R Br.
Osmunda cinnamomea L.

Lypopodiaceæ.
N. Lycopodium lucidulum Michx.
N. L. Selago L.
N. Selaginella rupestris Spring.
Isoetes melanopoda Gay.